SELECTED PAPERS FROM

ICNAAM 2007 *AND* ICCMSE 2007

To learn more about AIP Conference Proceedings,
including the Conference Proceedings Series, please visit the webpage
http://proceedings.aip.org/proceedings

SELECTED PAPERS FROM

ICNAAM 2007 AND ICCMSE 2007

Special Presentations at the

International Conference on Numerical Analysis and Applied Mathematics 2007

held in Corfu, Greece on 16 – 20 September 2007

and at the

International Conference on Computational Methods in Science and Engineering 2007

held in Corfu, Greece on 25 – 30 September 2007

EDITORS

Theodore E. Simos
University of Peloponnese
Tripolis, Greece

George Maroulis
University of Patras
Patras, Greece

CO-EDITORS

George Psihoyios
University of Buckingham
Buckingham, United Kingdom

Ch. Tsitouras
Technological Educational Institution of Chalkis
Psahna, Greece

All papers have been peer-reviewed.

SPONSORING ORGANIZATIONS
Greek Ministry of Education and Religious Affairs
European Society of Computational Methods
 in Sciences and Engineering (ESCMSE)

Melville, New York, 2008
AIP CONFERENCE PROCEEDINGS ■ 1046

Editor

Theodore E. Simos
Department of Computer Science and Technology
Faculty of Sciences and Technology
University of Peloponnese
GR-221 00 Tripolis
Greece
E-mails: tsimos@mail.ariadne-t.gr
        tsimos.conf@gmail.com

George Maroulis
Department of Chemistry
University of Patras
GR-26500 Patras
Greece
Email: maroulis@upatras.gr

Co-Editors
George Psihoyios
University of Buckingham
MK18 1EG Buckingham
United Kingdom
E-mail: g.psihoyios@ntlworld.com

Ch. Tsitouras
Department of Applied Sciences
Technological Educational Institution of Chalkis
School of Technolotical Applications (STEF)
GR34400 Psahna
Greece
E-mail: tsitoura@teihal.gr

L.C. Catalog Card No. 2008934701

ISBN 978-0-7354-0574-5
ISSN 0094-243X

Printed in the United States of America

CONTENTS

ICCMSE 2007

ICNAAM 2007

Preface: Special Presentations of the International Conference on Numerical Analysis and Applied Mathematics 2007 (ICNAAM-2007) and of the International Conference on Computational Methods in Sciences and Engineering 2007 (ICCMSE 2007)

ICNAAM 2007

Official Conference of the European Society of Computational Methods in Sciences and Engineering (ESCMSE)

The International Conference on Numerical Analysis and Applied Mathematics 2007 (ICNAAM-2007) has been scheduled to take place at Hotel Marbella in Corfu between 16th and 20th September 2007

The aim of the conference is to bring together leading scientists of the international Numerical and Applied Mathematics community and to attract original research papers of very high quality.

More than 400 papers have been submitted in order to be considered for presentation at ICNAAM-2007. From these submissions we have selected 212 papers after an international peer review by at least two independent reviewers.

We would like to thank:

- The eminent Scientific Committee of ICNAAM-2007 for their help and support (see the Conference Details in subsequent pages).

- The Symposia and Sessions organizers for their considerable effort, as well as for their valuable job in relation to the further development of the conference.

- The distinguished invited speakers for their difficult task and for their acceptance to give keynote lectures on their respective fields of expertise.

- The Organizing Committee (Dr. D.P. Sakas, Dr. Z.A. Anastassi, Mr. T.V. Triantafyllidis) for their valuable assistance. Special thanks to Dr. Z.A. Anastassi (Editorial Assistant of Professor Simos) for his help in typesetting of this Volume.

- Special thanks for the Secretary of ICNAAM-2007, Mrs Eleni Ralli-Simou (who is also the Administrative Secretary of the European Society of Computational Methods in Sciences and Engineering (ESCMSE)) for her excellent job.

- The Greek Ministry of Education for its financial support.

Professor Dr. T.E. Simos, Editor
President of the European Society of Computational Methods in Sciences and Engineering (ESCMSE), Active Member of the European Academy of Sciences and Arts; University of Peloponnese, Greece

Dr G. Psihoyios, Co-Editor
Vice-President of the European Society of Computational Methods in Sciences and Engineering (ESCMSE); Fellow of the IMA, UK

Professor Ch. Tsitouras, Co-Editor
Department of Applied Sciences, Technological Educational Institute of Chalkis, Greece

July 2008

ICCMSE 2007

Recognized Conference of the European Society of Computational Methods in Sciences and Engineering (ESCMSE)

The International Conference on Computational Methods in Sciences and Engineering 2007 (ICCMSE 2007) has been scheduled to take place at Hotel Marbella in Corfu between 25th and 30th September 2007

Topics of general interest are:

Computational Mathematics, Theoretical Physics and Theoretical Chemistry. Computational Engineering and Mechanics, Computational Biology and Medicine, Computational Geosciences and Meteorology, Computational Economics and Finance, Scientific Computation. High Performance Computing, Parallel and Distributed Computing, Visualization, Problem Solving Environments, Numerical Algorithms, Modelling and Simulation of Complex System, Web-based Simulation and Computing, Grid-based Simulation and Computing, Fuzzy Logic, Hybrid Computational Methods, Data Mining, Information Retrieval and Virtual Reality, Reliable Computing, Image Processing, Computational Science and Education etc.

The International Conference of Computational Methods in Sciences and Engineering (ICCMSE) is unique in its kind. It regroups original contributions from all fields of the traditional Sciences, Mathematics, Physics, Chemistry, Biology, Medicine and all branches of Engineering. It would be perhaps more appropriate to define the ICCMSE as a Conference on Computational Science and its applications to Science and Engineering. Based on the universality of mathematical reasoning the ICCMSE favors the interaction of various fields of Knowledge to the benefit of all. Emphasis is given on the multidisciplinary character of the Conference. The principal ambition of the ICCSME is to promote the exchange of novel ideas through the close interaction of research groups from all Sciences and Engineering.

In addition to the general programme the Conference offers an impressive number of Symposia. The purpose of this move is to define more sharply new directions of expansion and progress for Computational Science.

We note that for ICCMSE there is a co-sponsorship by American Chemical Society

More than 750 papers have been submitted for consideration for presentation in ICCMSE 2007. From these papers we have selected 362 papers after international peer review by at least two independent reviewers. These accepted papers have been presented at ICCMSE 2007.

We would like also to thank:

- The Scientific Committee of ICCMSE 2007 (see in page iv for the Conference Details) for their help and their important support. We must note here that it is a great honour for us that leaders on Computational Sciences and Engineering have accepted to participate in the Scientific Committee of ICCMSE 2007.

- The Organizers of the Symposia for their excellent editorial work and their efforts for the success of ICCMSE 2007.

- The highlighted lecturers and the invited speakers for their acceptance to give keynote lectures on Computational Sciences and Engineering.

- The Organizing Committee for their help and activities for the success of ICCMSE 2007.

- Special thanks for Dr. Zacharias Anastassi (Editorial Assistant of Professor Simos) for his help in typesetting of this Volume.

- Special thanks for the Secretary of ICCMSE 2007, Mrs Eleni Ralli-Simou (which is also the Administrative Secretary of the European Society of Computational Methods in Sciences and Engineering (ESCMSE)) for her excellent job.

Prof. Theodore E. Simos, Academician of EAS, EASA, EAASH
President of ESCMSE
Chairman ICCMSE 2007
Editor of the Proceedings
Department of Computer Science
and Technology
University of the Peloponnese
Tripolis
Greece

Prof. George Maroulis
Co-Editor of the Proceedings
Chairman ICCMSE 2007
Department of Chemistry
University of Patras
Patras
Greece

November 2007

CONFERENCES DETAILS

ICNAAM 2007

International Conference of Numerical Analysis and Applied Mathematics 2007 (ICNAAM 2007), Hotel Marbella (Agios Ioannis Peristeron, Corfu), Corfu, Greece, 16-20 September 2007.

Official Conference of the European Society of Computational Methods in Sciences and Engineering (ESCMSE)

Chairman and Organiser

Professor T.E. Simos, President of the European Society of Computational. Methods in Sciences and Engineering (ESCMSE). Member of the Presidium of the European Academy of Sciences (Editor of the Annals of the European Academy of Sciences (Applied Mathematics) - http://www.eurasc.org/edi_board.asp). Active Member of the European Academy of Sciences and Arts, Corresponding Member of the European Academy of Sciences, Corresponding Member of European Academy of Arts, Sciences and Humanities, Department of Computer Science and Technology, Faculty of Sciences and Technology, University of Peloponnese, Greece.

Co-Chairmen

Dr. G. Psihoyios, Vice-President of the European Society of Computational Methods in Sciences and Engineering (ESCMSE), Fellow of the IMA, University of Buckingham, UK

Professor Ch. Tsitouras, Department of Applied Sciences, Technological Educational Institute of Chalkis, Greece .

Scientific Committee

- **Professor G. vanden Berghe**, Belgium
- **Professor Peter Bjørstad**, Norway
- **Professor S.C.Brenner**, USA
- **Professor J. R. Cash**, UK
- **Professor R. Cools**, Belgium
- **Professor A. Cuyt**, Belgium
- **Professor B. Fischer**, Germany
- **Professor R. W. Freund**, USA
- **Professor I. Gladwell**, USA
- **Professor Gene Golub**, USA
- **Professor B. Hendrickson**, USA
- **Professor Axel Klar**, Germany
- **Dr. W.F. Mitchell**, USA
- **Dr. G. Psihoyios**, UK
- **Professor T.E. Simos**, Greece
- **Professor W. Prößig**, Germany

- **Professor Francis Sullivan**, USA
- **Dr. Ch. Tsitouras**, Greece
- **Professor G. Alistair Watson**, UK

Invited Speakers

- **Professor Dr. C. W. de Boor (Carl),** Emeritus Professor, Department of Computer Sciences and Department of Mathematics, University of Wisconsin - Madison, USA

- **Professor Dr. C. W. Gear (Bill),** Senior Scientists, Chemical Engineering, Princeton University (zero-time appointment), Emeritus President, NEC Research Institute, Emeritus Professor, Department of Computer Science, University of Illinois at Urbana-Champaign, USA

- **Professor Dr. Mariano Gasca**, Depto. Matemã¡tica Aplicada, Fac. Ciencias, Universidad de Zaragoza, 50009 Zaragoza, Spain

- **Professor Dr. G Alistair Watson**, University of Dundee, Division of Mathematics, Dundee DD1 4HN, Scotland.

- **Professor Dr. Eugene Tyrtyshnikov**, Deputy Director, Institute of Numerical Mathematics, Russian Academy of Sciences, Gubkina Street, 8, Moscow 119333, Russia

- **Professor Dr. Karel in 't Hout**, Department of Mathematics and Computer Science, University of Antwerp, Campus Middelheim Middelheimlaan 1, B-2020 Antwerp, Belgium

- **Professor Dr. Jaime Keller**, Theoretical Physics and Chemistry, Center for Computational Materials Science, E134 Institut für Allgemeine Physik, Technische Universität Wien and Departamento de Física y Química Teórica, Facultad de Química, Universidad Nacional Autónoma de México

- **Professor Ezio Venturino**, Dipartimento di Mathematica, Universita' di Torino, via Carlo Alberto 10, 10123 Torino, Italy

- **Professor Dr. Yaroslav D. Sergeyev**, Dipartimento di Elettronica, Informatica e Sistemistica, Università della Calabria, Via P. Bucci, Cubo 41-C87036 Rende (CS), Italia and Professor (part-time contract) University of Nizhni Novgorod, pr. Gagarina, 23, 603600 Nizhni Novgorod, Russia

ICCMSE 2007

The International Conference on Computational Methods in Sciences and Engineering 2007 (ICCMSE 2007) has been scheduled to take place at Hotel Marbella in Corfu between 25th and 30th September 2007

Recognized Conference of the European Society of Computational Methods in Sciences and Engineering (ESCMSE)

Chairmen and Organizers

Professor Theodore E. Simos, Academician of EAS, EASA, EAASH, Member of the Presidium of EAS, President of ESCMSE, Department of Computer Science and Technology, University of the Peloponnese, Tripolis

Professor George Maroulis, Department of Chemistry, University of Patras, Patras, Greece

Scientific Committee

Dr. B. Champagne, Université de Namur, Belgique
Prof. S. Farantos, University of Crete, Greece
Prof. I. Gutman, University of Kragujevac, Serbia
Prof. P.Mezey, Memorial University of Newfoundland , Canada
Prof. C. Pouchan, Université de Pau, France
Dr. G. Psihoyios, Vice-President ESCMSE
Prof. B. M. Rode, University of Innsbruck, Austria
Prof. A. J. Thakkar, University of New Brunswick, Canada

Highlighted Lectures:

Reinhart Ahlrichs, University of Karlsruhe
N.Y. Öhrn, University of Florida
Henry F. Schaefer III, University of Georgia

Highlighted Keynote Lecture

E. N. Economou, University of Crete, Greece

Invited Speakers

Lorenz S. Cederbaum, Germany
Marcus Elstner, Germany
Patrick W. Fowler, UK
Pavel HOBZA, Czech Republic
Shashi KARNA, USA
Christoph H. Keitel, Germany
Klaus Lucas, Germany
J.P.Malrieu, France
Debashis Mukherjee, INDIA
Kazuaki J. Murakami, Japan
Cleanthes A. Nicolaides, Greece

Koji OHTA, Japan
Dennis Salahub, Canada
Feng Wang, Australia
Francesco Zerbetto, Italy

Keynote Speakers

Xavier Assfeld, France
Jochen Autschbach, USA
T. Bancewicz, Poland
Cristian V. Ciobanu, USA
Patrick Norman, Sweden
Alejandro Toro-Labbé, Chile
Patrizia Calaminici, Mexico
Yuriko Aoki, Japan
Jorge J. Kohanoff, Northern Ireland (UK)
Shiro Koseki, Japan
Jorge Alberto Morales, USA
Joe McDouall, UK
Leticia Gonzalez, Germany
Manfred Lein, Germany
Sourav Pal, India
M. G. Papadopoulos, Greece
Sylvain Picaud, France
S. RAMASESHA, India
Antonio RIZZO, Italy
Luis Serrano-Andrés, Spain
Hideo Sekino, Japan
Holger Vach, France
Yasunori YOSHIOKA, Japan
Claudio Zannoni, Italy

Organizing Committee

Mrs Eleni Ralli-Simou (Secretary of ICCMSE 2007)
Dr. Z. A. Anastassi
Mr. T. V. Triantafyllidis
Mr. G. Vourganas
Dr. Th. Monovasilis
Mr. G. Panopoulos
Mr. Nektarios Tselios

European Society of Computational Methods in Sciences and Engineering (ESCMSE)

AIMS AND SCOPE

The *European Society of Computational Methods in Sciences and Engineering (ESCMSE)* is a non-profit organization. The URL address is: http://www.uop.gr/escmse/

The aims and scopes of *ESCMSE* is the construction, development and analysis of computational, numerical and mathematical methods and their application in the sciences and engineering.

In order to achieve this, the *ESCMSE* pursues the following activities:

• Research cooperation between scientists in the above subject.
• Foundation, development and organization of national and international conferences, workshops, seminars, schools, symposiums.
• Special issues of scientific journals.
• Dissemination of the research results.
• Participation and possible representation of Greece and the European Union at the events and activities of international scientific organizations on the same or similar subject.
• Collection of reference material relative to the aims and scope of *ESCMSE*.

Based on the above activities, *ESCMSE* has already developed an international scientific journal called:

Journal of Numerical Analysis, Industrial and Applied Mathematics (JNAIAM). This is a journal under the copyright of ESCMSE. This is the journal for original research papers, review papers and letters. We note that this journal is electronically free for the academic community. URL address: **http://www.jnaiam.org**

CATEGORIES OF MEMBERSHIP

European Society of Computational Methods in Sciences and Engineering (ESCMSE)

Initially the categories of membership will be:

• **Full Member (MESCMSE):** PhD graduates (or equivalent) in computational or numerical or mathematical methods with applications in sciences and engineering, or others who have contributed to the advancement of computational or numerical or mathematical methods with applications in sciences and engineering through research or education. Full Members may use the title MESCMSE.

• **Associate Member (AMESCMSE):** Educators, or others, such as distinguished amateur scientists, who have demonstrated dedication to the advancement of computational or numerical or mathematical methods with applications in sciences and engineering may be elected as Associate Members. Associate Members may use the title AMESCMSE.

• **Student Member (SMESCMSE):** Undergraduate or graduate students working towards a degree in computational or numerical or mathematical methods with applications in sciences and engineering or a related subject may be elected as Student Members as long as they remain students. The Student Members may use the title SMESCMSE

• **Corporate Member:** Any registered company, institution, association or other organization may apply to become a Corporate Member of the Society.

REMARKS

1. After three years of full membership of the European Society of Computational Methods in Sciences and Engineering, members can request promotion to Fellow of the European Society of Computational Methods in Sciences and Engineering. The election is based on international peer-review. After the election of the initial Fellows of the European Society of Computational Methods in Sciences and Engineering, another requirement for the election to the Category of Fellow will be the nomination of the applicant by at least two (2) Fellows of the European Society of Computational Methods in Sciences and Engineering.

2. All grades of members other than Students are entitled to vote in Society ballots.

3. All grades of membership other than Student Members receive the official journal of the ESCMSE.

We invite you to become part of this exciting new international project and participate in the promotion and exchange of ideas in your field.

ICCMSE 2007

Localized Multi-Reference Approach
for Mixed-Valence Systems

W. Helal, S. Evangelisti, T. Leininger and D. Maynau

Laboratoire de Chimie et Physique Quantique, UMR 5626 du CNRS, IRSAMC,
Université Paul Sabatier, 118, Route de Narbonne, 31062 Toulouse, France

Abstract. The electronic structure and some important intra-molecular charge transfer parameters were investigated at CAS-SCF, MRCI, CAS+S and multi-reference localization levels of theory for purely organic mixed-valence molecules. In particular, a spiro cation has been taken as a model system. The potential energy surfaces of the ground and the lower three excited electronic states have been computed within a two-state model, at CAS-SCF using TZP basis for the spiro cation, and an adiabatic double-well potential has been obtained for the ground electronic state. Our analysis of the geometry through the reaction coordinate indicate that the spiro cation is a valence trapped bistable system. The effect of non-dynamical correlation, using a localized orbital approach, was found to be crucial for a quantitative description of the electronic structure and some important electron transfer parameters of these organic mixed-valence systems.

Keywords: multi-reference methods, localization, CAS-SCF, mixed-valence molecules, spiro, bis(triarylamine)
PACS: 31.25.Qm, 31.50.Bf, 31.50.Df

INTRODUCTION

Mixed-valence compounds are a special class of intramolecular charge transfer molecules with two or more redox sites existing in different oxidation states. The importance of these fascinating molecular architectures are their use as model compounds for the fundamental study of the electron transfer process as well as there wide range applications in molecular devices. Robin and Day classified mixed-valence systems into three types: class I, completely valence trapped with no electronic coupling between redox sites; class II, valence trapped (weak coupling); and class III, delocalized valency (strong coupling).

In this contribution, a CAS-SCF with different active spaces and subsequent MRCI and orbital localization methods have been applied to a model molecular cation: 5,5'(4H,4H')-spirobi[cyclopenta[c]pyrrole]2,2',6,6'tetrahydro cation, 1^+, this model molecular cation, see Fig. 1, is useful since its relatively small molecular size permits high-level *ab-initio* investigations. Mixed-valence systems shows different charge localization extents, coupling strengths, electron transfer (ET) potential energy barriers and other ET parameters. It is not the objective of this contribution to give exact quantitative values of these parameters, however, we will try to answer some other important questions like: To what molecular size limit we can apply a high level multi-configurational ab-initio method? and, what is the advantage of using such a method for a quantitative description of the electronic structure of mixed valence compounds? We will show that taking the advantage of the locality of the important "active sites" in a relatively large molecular system, one can reduce substantially the computational cost of the non-dynamical correlation and subsequent dynamical correlation correction calculations. Moreover, it will be shown that the use of the non-dynamical correlation for mixed valence molecular systems is very appealing if accurate quantitative results of the electronic states involved in the electron transfer process are desired.

In the next section, the spiro molecular system and the methods used are briefly described. The last section will be devoted to the results obtained with some conclusions and perspectives for future work.

SPIRO MOLECULE AND THE THEORETICAL MODELS USED

The spiro molecular cation 1^+ consists of two equivalent π moieties, lying onto two orthogonal planes, and separated by a rigid σ bridge. If an electron is extracted from this molecule, the resulting hole tends to localize either on the left or the right π system, inducing a deformation of the molecular geometry. Therefore two equivalent minima exist for the cation, and the ground state presents the double-well surface which is typical for mixed-valence systems of class

CP1046, *Selected Papers from ICNAAM 2007 and ICCMSE 2007*
edited by T. E. Simos, G. Maroulis, G. Psihoyios, and Ch. Tsitouras
© 2008 American Institute of Physics 978-0-7354-0574-5/08/$23.00

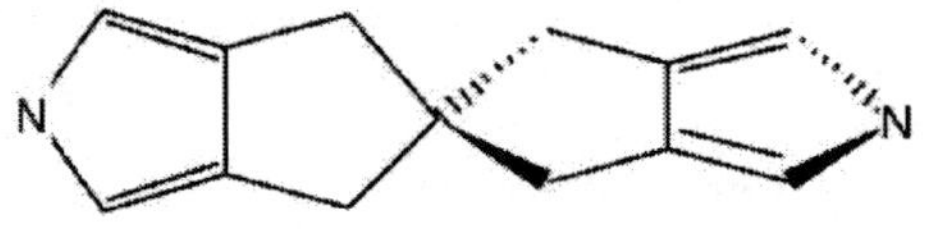

1

FIGURE 1. Spiro molecule **1**

II. A classical two-state model is adopted for the intramolecular charge transfer between the two moieties in the Spiro cation.

The Complete Active Space (CAS) was taken to be all or part of the valence π electrons. We have considered CAS(7/4), CAS(7/8) and CAS(11/10) for $\mathbf{1}^+$, the last active space contain, in principle, the complete valence π system in the cation [1].

We have used a method, recently developed by our group, for obtaining localized orbitals of CAS-SCF type [2]. One of the main characteristics of the method is that the final localized orbitals are directly obtained and optimized from guess localized orbitals. In such a case, one may choose, among the large number of possible CAS spaces, the particular set of active orbitals that are relevant for the study of a particular phenomena. Furthermore, dynamical correlation can be introduced to the multi-configurational wavefunction in order to correct the latter, for instance, in this study, CAS+S calculations were computed for the molecular cations starting from localized guess orbitals. Another relevant method that is also developed by the Toulouse group is the selected CI approach [3]. In this scheme, only the excitations involving an electron transfer between close orbitals are retained. Two-electron excitations are then produced as the combination of acceptable one-electron excitations. Some preliminary tests that have been performed show a slow growth of the CI size as a function of the system size, at the same time, the energy differences show a remarkable agreement with full CI-SD results.

RESULTS AND DISCUSSION

The reaction coordinate of $\mathbf{1}^+$ was obtained by mixing linearly the optimized coordinates of the two equilibrium geometries:

$$Q(\xi) = \left(\xi - \frac{1}{2}\right) Q_B + \left(\xi + \frac{1}{2}\right) Q_A \tag{1}$$

where $Q(\xi)$ is the nuclear configuration at the point ξ on the reaction path, while Q_A and Q_B represent the nuclear coordinates of the two optimized charge localized geometries. The mixing parameter ξ was varied from -1.50 to $+1.50$ in steps of 0.05. In such a way, Q_A and Q_B are the geometries of the two minima, and the internal coordinates of spiro cation at the crossing seam, where the charge is delocalized over the molecular two moieties, is calculated as the average of the two charge localized geometries, and it is obtained at the crossing seam point $\xi = 0.00$. Geometry optimization of the equilibrium coordinates of $\mathbf{1}^+$ were performed at ROHF/TZP level. The Born-Oppenheimer potential energy surfaces were obtained by calculating the energies of lower electronic states at each step value of the parameter ξ, using CAS(7/4)/TZP and MRCI(SD)/TZP. Fig. 2 shows the potential energy surface of $\mathbf{1}^+$ at CAS(7/4)/TZP. In order to test the quality of the averaged geometry at the crossing seam, the electronic state energies of the latter were compared with the corresponding state energies of the optimized geometry. The results show a very small difference between these electronic states In addition, a comparison of the corresponding bond lengths of the Spiro cation at the averaged and optimized geometries at the crossing seam also shows an extremely small difference between the two structures [1].

The geometry of the charge bearing moiety of the radical cation in the case of the charge localized state was found to be very close to that of the di-cation. On the other hand, the geometry of the neutral moiety of the radical cation was very close to that of the neutral spiro molecule. This leads us to conclude that the spiro cation is a valence-trapped mixed bistable system of class II.

The energy difference between the CAS-SCF and CAS+S using different active spaces for the ET energy barrier was found to be around 50% lower for the case of CAS+S starting from localized guess orbitals. Moreover, $2V_{ab}$ computed value, where V_{ab} is the ET matrix element, was found to be decreased by around 10% when performing

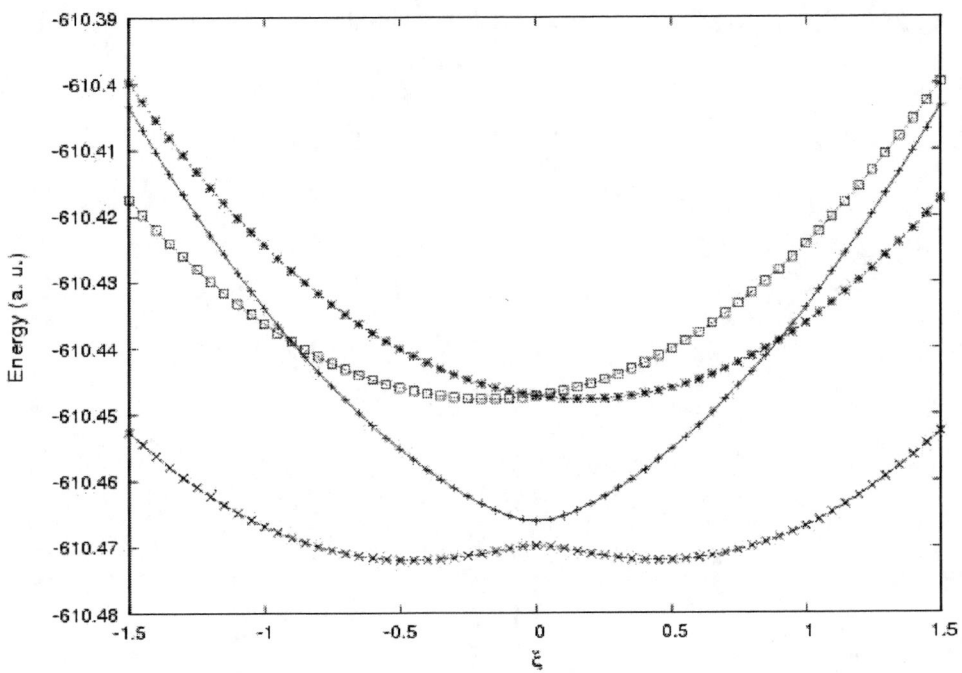

FIGURE 2. Potential energy surfaces of the spiro cation ground state and other low-lying excited states calculated at CAS(7/4)/TZP

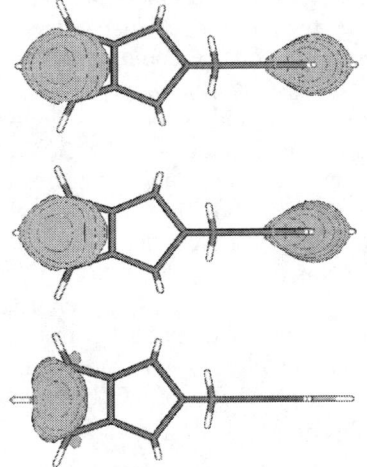

FIGURE 3. delocalized a_1 orbitals (top and middle) compared to a localized a_1 orbital (bottom)

CAS+S calculation. Fig. 3 shows two delocalized a_1 orbitals (top and middle) a and a local a_1 symmetry orbital (bottom) calculated by the Toulouse group algorithm CASDI.

The preliminary results of local MR-CI methods tested on 1^+ produced accurate description of the electronic interaction and produced the double-minimum surface, this encouraged us to extend our study on significantly larger molecules. In particular, the electronic structure of bis(triarylamine) cations 2^+ and 3^+, see Fig. 4, is currently under investigation. The two bis(triarylamine) cations N,N,N',N'-Tetra(4-methoxyphenyl)-1,4-phenylenediamine cation, 2^+; and Bis4-[N,N-di(4-methoxyphenyl)amino]phenylbutadiyne cation, 3^+. Bis(triarylamine) **2** and **3** are composed of two aromatic triarylamine units separated by a highly conjugated bridge between the two redox sites. The dihedral

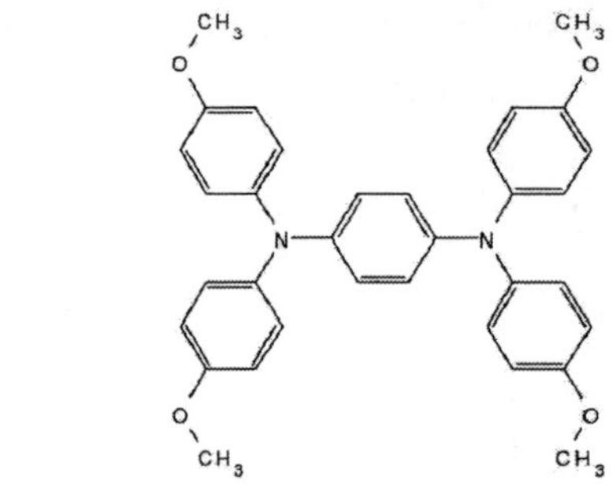

FIGURE 4. Bis(triarylamine) neutral molecules **2** and **3**

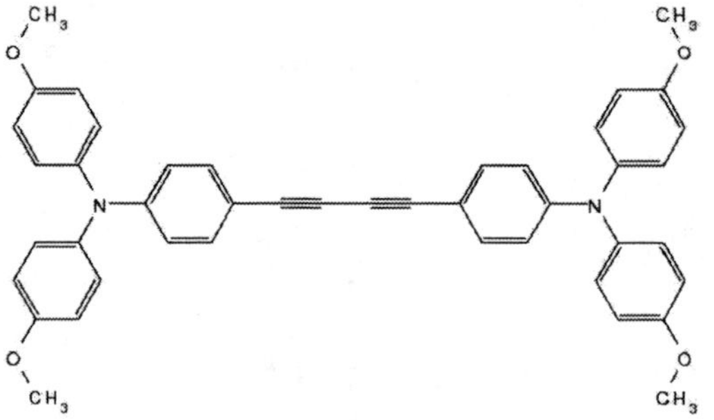

angles of the triarylamine planes of these cations are not planer nor orthogonal to each others and these angles may vary depending on the distance between the two charge bearing centers. One main difference between **1**$^+$ from one side and **2**$^+$ and **3**$^+$ from the other side is that the charge could be found to be localized in the bridge[1] in the case of the latter, thus, using a two-state model for the bis(triarylamine) cations **2**$^+$ and **3**$^+$ is questioned. for each of these two molecules, the π orbitals of the two nitrogen atoms and the bridge were used to construct the active space.

In order to perform MR-CI calculations in a localized context for **2**$^+$ and **3**$^+$, local orbitals should be obtained, as a first step, in order to be able to perform a selective treatment of the correlation in the different regions of the molecule. Then, the electrons located on the outer aryl cycles, those bearing the methoxy group, are kept frozen at the SCF level [4]. In this way, the steric and polarization effects of these groups on the nitrogen centers are fully taken into accounts, while the size of the correlated region remains relatively small. Finally, the electrons that are located on the bridge orbitals are correlated via the excitation-selected method [3]. The preliminary results are very promising both in terms of the physical content of the problem and in the reduction of the correlation computational cost. It should be stressed that, up to now, theoretical investigations on such large systems are mainly performed at a semi-empirical level, or at most by using Density Functional Theory. The possibility of a fully ab initio treatment on mixed valence systems can open the way to a huge number of applications.

REFERENCES

1. W. Helal, S. Evangelisti, T. Leininger and D. Maynau, *J. Comput. Chem.* submitted.
2. D. Maynau, S. Evangelisti, N. Guihéry, C. J. Calzado, and J.-P. Malrieu, *J. Chem. Phys.*, **116**, 10060-10068 (2002).
3. B. Bories, D. Maynau and M.-L. Bonnet, *J. Comput. Chem.* **28**, 632-643, (2007).
4. J. Pitarch-Ruiz, C. J. Calzado, S. Evangelisti, and D. Maynau, *Int. J. Quant. Chem.* **106**, 609-622 (2006).

---

[1] In a hopping charge transfer model.

Toward Accurate Modelling of Enzymatic Reactions: All Electron Quantum Chemical Analysis combined with QM/MM Calculation of Chorismate Mutase

Toyokazu Ishida

Research Institute for Computational Science (RICS), National Institute of Advanced Industrial Science and Technology (AIST), Tsukuba Central 2, 1-1-1 Umezono, Tsukuba, 305-8568, Japan.

Abstract. To further understand the catalytic role of the protein environment in the enzymatic process, the author has analyzed the reaction mechanism of the Claisen rearrangement of *Bacillus subtilis* chorismate mutase (BsCM). By introducing a new computational strategy that combines all-electron QM calculations with *ab initio* QM/MM modelings, it was possible to simulate the molecular interactions between the substrate and the protein environment. The electrostatic nature of the transition state stabilization was characterized by performing all-electron QM calculations based on the fragment molecular orbital technique for the entire enzyme.

Keywords: *ab initio* QM/MM modeling, fragment molecular orbital method, enzyme reaction, chorismate mutase
PACS: 82.39.Rt Reactions in complex biological systems

INTRODUCTION

Recent progress in chemical theory and computational methods has opened the way for computer simulations of the structure-function relationship in biomolecules[1]. Given the wide variety of biological functions found in nature, understanding the catalytic power of enzymes is one of the biggest challenges of modern theoretical chemistry[2]. Although the general concept of transition state stabilization (TSS) is well recognized in many enzymatic processes, the atomic level detail of the TSS process remains a mystery. To gain further understanding of this process, the author has investigated the reaction profile of the Claisen rearrangement of *Bacillus subtilis* chorismate mutases (BsCM) by means of a new computational strategy that combines all-electron quantum chemical calculations with quantum mechanical/ molecular mechanical (QM/MM) modelling techniques.

Chorismate mutases are the enzymes that catalyze the conversion of chorismate into prephenate in the shikimate pathway, known as the biosynthetic route of aromatic amino acids in plants and microorganisms. The enzymatic reaction is considered to proceed via a chair-like pericyclic transition state with concerted C-O cleavage / C-C bond formation. Structural and biochemical investigations suggest that the highly polarized TS is stabilized through the electrostatic interaction of positively charged residue (Arg or Lys) adjacent to the ether oxygen of the substrate[3, 4, 5, 6]. To clarify the catalytic role of the protein environment many types of QM/MM and related calculations have been performed[7, 8, 9, 10, 11, 12, 13, 14, 15, 16, 17]. One research direction of QM/MM studies has been to analyze the molecular interaction between selected residue clusters and the substrate based on reaction path modelling[7, 8, 9, 10, 11, 12, 17]. Although some calculations supported the electrostatic stabilization hypothesis, other results suggest that the conformational restriction of substrate is important[14, 15, 16]. At present, any fundamental approach used to describe enzyme reactions is limited to *ab initio* QM/MM calculations at best, where large protein structures are described using empirical force fields. To increase further the capability of the standard QM/MM method, a computational strategy combining the all-electron QM method with *ab initio* QM/MM modelling was developed. In particular, the fragment molecular orbital (FMO) technique was used as the basis for the all-electron QM methodology[18, 19]. Using this combined method the electrostatic stabilization factor in the enzymatic process and the reaction profile of wild-type BsCM, as well as that of several mutants, was determined. By thoroughly comparing several reaction profiles and by performing the interaction energy component analysis from the quantum chemical viewpoint, the catalytic role of the protein environment is examined.

Due, primarily, to a lack of computational resources for performing full QM reaction path searches, the following procedure was implemented: (1) the reaction path and the energy profile of the wild-type BsCM were first calculated using *ab initio* QM/MM modelling; (2) ES, TS and product geometries of several mutant enzymes were determined based on the wild-type reaction profile; (3) the all-electron QM analyses based on the fragment molecular orbital technique was performed for the important geometries along the reaction pathway.

METHOD AND COMPUTATION

QM/MM modelling: A standard QM/MM partition scheme was employed. The interactions between the QM and MM regions are defined as,

$$\hat{H}_{QM/MM} = \hat{H}^{elec}_{QM/MM} + \hat{H}^{vdw}_{QM/MM} + \hat{H}^{strain}_{QM/MM} \tag{1}$$

where the first term is evaluated from the electronic structure calculations, while the other two terms are estimated from the classical force field. Because the main purpose of QM/MM modelling is to construct a reasonable reaction profile, the QM region was limited to the substrate for simplicity. In this case, a no strain term linking the QM-MM boundaries was needed. The Restricted Hartree-Fock (RHF) method (with a 6-31(+)G** basis set, where diffuse functions were only added to the two carboxylic groups), and the AMBER parameter set (parm.96) for force field calculations were used[20]. All QM/MM calculations were performed using the author's original code, reported in previous work[21].

The initial coordinates of the enzyme were obtained from the X-ray crystal structure of BsCM bound to the endo-oxabicyclic transition state analogue (TSA), PDB code 2CHT. Assuming that the standard physiological descriptions of enzyme processes holds, the ionized state of the polar residue was determined, and hydrogen atoms were added to the initial model. The reactant chorismate was placed at the original X-ray coordinate of TSA. After locating the TS geometry from the QM/MM treatment, the minimum energy reaction path of the wild-type reaction was calculated. All stationary structures were confirmed using vibrational frequency calculations. The mutant enzyme structures in the initial models were constructed using the wild-type structure. Each structure was optimized by overlapping the heavy atom coordinates of the mutation points on to that of wild-type geometry.

All-Electron Calculations using the Fragment Molecular Orbital Method: The FMO method employs the *ab initio* MO treatment to obtain an estimate of the electronic structure of large molecular systems[18]. If a system is separated into small discrete fragments, and the *ab initio* MO calculations are performed for all fragment pairs using the surrounding electrostatic potential (ESP) due to the remaining fragments, the total energy of the system is evaluated as,

$$E_{total} = \sum_I E_I + \sum_{I>J} (E_{IJ} - E_I - E_J) \tag{2}$$

where E_I and E_{IJ} are the monomer and dimer energies, respectively. The monomer and dimer total energies include the environmental ESP from the surrounding monomers. When the ESP term is subtracted from the E_I and E_{IJ}, eq(2) can be rewritten as a sum of the modified monomer and the interfragment interaction energies ΔE_{IJ},

$$E_{total} = \sum_I E'_I + \sum_{I>J} \Delta E_{IJ} \tag{3}$$

The second term in eq(3), ΔE_{IJ}, is the *effective pair interaction energy* that includes the many-body effects through the external field of the system. The analysis of the energy component of ΔE_{IJ} provides the detail of the environmental effect in the enzymatic process. It is important to note that in contrast to the standard QM/MM model, which assigns a fixed point charge at each atomic site, the FMO method can provide the charge transfer and the mutual polarization effect due to the self-consistent electrostatic environment.

All FMO calculations were performed using the GAMESS quantum chemical software package[22]. The protein is divided into fragments at the C_α position to keep the peptide bonds intact. One residue per fragment partition was employed for the ease of estimating the energetic component of particular catalytic residue elements. The hybrid sp^3 orbitals of the carbon atoms were used to properly divide the molecular orbital space at the bond fraction points. A 6-31(+)G* basis set was used, where diffuse functions were added to all carboxyl groups (Asp, Glu, C-Term) in the enzyme and the substrate. To accurately describe the molecular interaction between the substrate and the enzyme, a proper treatment of the electron correlation is necessary. In this work all intermolecular interactions between the substrate and the surrounding amino acid residue were evaluated using the MP2 level of the FMO framework[19]. Further technical details are described in the literature[17].

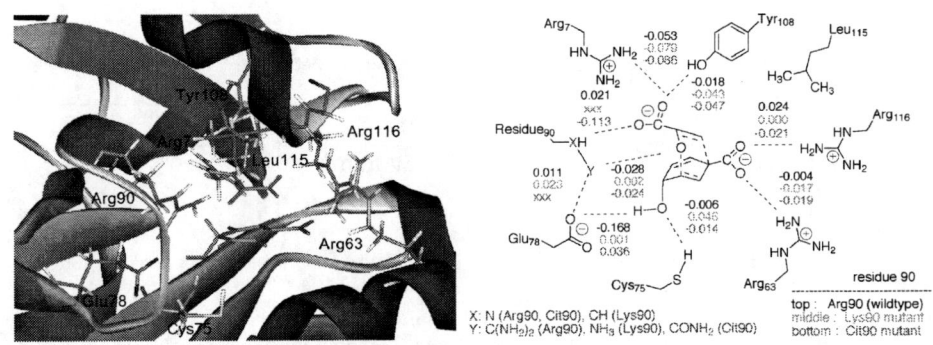

FIGURE 1. (left) The QM/MM optimized TS geometry of the wild-type enzyme. (right) The difference of hydrogen bonding distances (TS minus ES) in each reaction system. Important hydrogen bond parameters for three types of enzymes (wild-type in black, lys90 mutant in green, cit90 mutant in purple) are given in Å. The side chain atoms of residue90 are represented by ' X ' and ' Y '.

RESULTS AND DISCUSSION

The schematic picture of the hydrogen bonding network and the 3D model of an active site in the TS geometry are presented in Figure 1. Only the important residue elements are included in this figure: Arg63, Cys75 (in domain1), Arg7, Glu78, Arg90, Tyr108, Leu115 and Arg116 (in domain2). In TS the limit of the C-O bond length before breaking is 2.130 Å, while that of forming C-C is 2.546 Å. This result indicates that the isomerization proceeds in an asynchronous fashion. In addition the location of TS is slightly shifted to the reactant side. From the analysis of the molecular geometries along the reaction coordinate, large conformational changes in and around the active site were not found. In the case of the wild-type reaction, the major structural change is a rearrangement of the hydrogen bonding network in the Glu78-Arg90-substrate alignment, while other residue elements at the active site (Arg63, Arg116, Arg7 and Tyr108) work to fix the relative orientation of the substrate.

Figure 2 is a plot of the interaction energy decomposition into each residue contribution. The result was evaluated from ΔE_{IJ} in eq(3) by performing all-electron QM calculations for the entire enzyme. It is apparent that only a limited number of residue elements are responsible for the catalysis. Of these elements Arg90 is the most crucial contribution in the wild-type reaction. Furthermore, Glu78 and Arg7 exhibit a large attractive interaction. Of note, all polar residue elements that exhibit a strong interaction are found to be located inside the binding pocket; Arg90 interacts with the ether oxygen and the carboxylate group of the enol-pyruvyl side chain of the substrate, Glu78 is bound to the substrate hydroxyl group, and Arg7 forms hydrogen bonds with the buried carboxylate part. As the reaction proceeds, a slight movement of the C-terminal helix (Glu110~Arg116) is observed along the reaction pathway. In accordance with the structural change of the enol-pyruvyl group, the C-term helix closely approaches the substrate during the TS formation process. The large catalytic contribution of non-polar Leu115, located on the lid position of the catalytic pocket, originates from this structural movement.

The theoretical modelling of the wild-type reaction clearly demonstrated that the most crucial residue is Arg90, located on the domain 2 region. To identify the catalytic role of Arg90, mutant enzymes were prepared by replacing the Arg90 in the wild-type structure with Lys90 and Cit90 (citrulline, isosteric-neutral arginine analogue), and the same systematic analyses was repeated. The active site geometries of both the Lys90 and Cit90 variants are included in Figure 1. In both cases, no large conformational change is observed in the overall protein structure. However, changes were observed in the geometries around the mutation point. Due to a loss of the steric hindrance by the guanidium side chain of Arg90, most of the hydrogen bonds at the active site were slightly weakened in the ES complex. As a result the substrate forms a strong hydrogen bond with Glu78 and the ES, compared with the wild-type reaction, is further stabilized in the mutants. A slight modification of the geometrical parameters significantly influences the relative stability of the substrate. In particular, a sterical factor of Arg90 that bridges Glu78 to the substrate through hydrogen bonds seems to be a key element in achieving an effective catalysis.

A description of the molecular interaction between the substrate and the protein was obtained from the all-electron QM simulations. In the Lys90 mutant, although the charge effect that polarizes the substrate is similar to that of Arg90, a large catalytic loss is observed at the Lys90 position. The main reason is that ES is further stabilized because of the loss of the steric factor that prevents Glu78 to form a strong hydrogen bond with the substrate. In the case of the Cit90

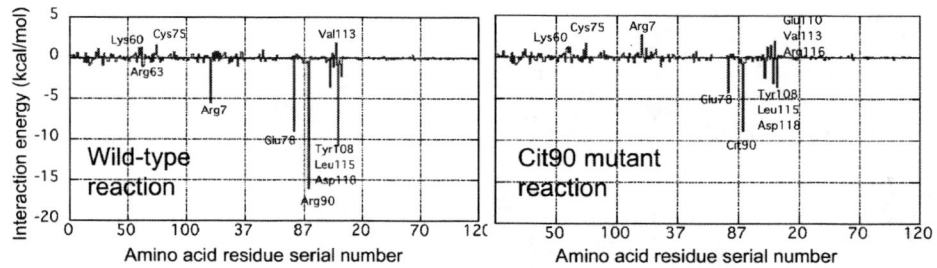

FIGURE 2. A plot of the interaction energy decomposition of the wild-type reaction (left) and the Cit90 mutant reaction (right). The x-axis is the amino residue serial number and the y-axis is the interaction energy difference ΔE_{IJ} between the TS and ES states (in kcal/mol).

mutant, structural changes break the fine balance in the hydrogen bonding of the Glu78-Arg90-substrate alignment. The charge distribution of the substrate is also different, compared to the other two cases, and a relatively large dipole is found in the ES complex. As a result, even in ES, the buried carboxylate part of the substrate forms strong hydrogen bonds with Arg7. The catalytic contribution of Cit90 is mainly due to the strong attraction to the buried carboxylate⁻ of the substrate, not the induced charge on the ether oxygen at the TS region. As is clearly shown in Figure 2, Cit90 stabilizes the TS in a different way than the Arg90 in the wild-type reaction.

ACKNOWLEDGMENTS

The author is grateful to Dr. Kazuo Kitaura for his continuing encouragement. This work was supported by the Next Generation Super Computing Project, Nanoscience Program, MEXT, Japan. Some of the numerical calculations were performed at the Computer Center of the Institute for Molecular Science (IMS).

REFERENCES

1. Special Issue on *Molecular Dynamics Simulations of Biomolecules, Acc. Chem. Res.* **35**, vol 6, (2002).
2. Special Issue on *Principles of Enzyme Catalysis, Chem. Rev.* **106**, vol 8, (2006).
3. Y. M. Chook, H. Ke and W. N. Lipscomb, *Proc. Natl. Acad. Sci. U.S.A.* **90**, 8600-8603 (1993).
4. D. J. Gustin, P. Mattei, P. Kast, O. Wiest, L. Lee, W. W. Cleland, and D. Hilvert, *J. Am. Chem. Soc.* **121**, 1756-1757 (1999).
5. P. Kast, C. Grisostomi, I. A. Chen, S. Li, U. Krengel, Y. Xue, and D. Hilvert, *J. Biol. Chem.* **275**, 36832-36838 (2000).
6. A. Kienhöfer, P. Kast and D. Hilvert, *J. Am. Chem. Soc.* **125**, 3206-3207 (2003).
7. P. D. Lyne, A. J. Mulholland, and W. G. Richards, *J. Am. Chem. Soc.* **117**, 11345-11350 (1995).
8. S. Martí, J. Andrés, V. Moliner, I. Silla, I. Tuñón, J. Bertrán. and M. J. Field, *J. Am. Chem. Soc.* **123**, 1709-1712 (2001).
9. Y. S. Lee, S. E. Worthington, M. Krauss, and B. R. Brooks, *J. Phys. Chem. (B)* **106**, 12059-12065 (2002).
10. B. Szefczyk, A. J. Mulholland, K. E. Ranaghan, and W. A. Sokalski, *J. Am. Chem. Soc.* **126**, 16148-16159 (2004).
11. A. Crespo, D. A. Scherlis, M. A. Martí, P. Ordejón, A. E. Roitberg, and D. A. Estrin, *J. Phys. Chem. (B)* **107**, 13728-13736 (2003).
12. H. L. Woodcock, M. Hodošček, P. Sherwood, Y. S. Lee, H. F. Schaefer III, and Brooks, B. R. *Theor. Chem. Acc.* **109**, 140-148 (2003).
13. M. Štrajbl, A. Shurki, M. Kato, and A. Warshel, *J. Am. Chem. Soc.* **125**, 10228-10237 (2003).
14. S. Hur and T. C. Bruice, *J. Am. Chem. Soc.* **125**, 5964-5972 (2003).
15. H. Guo, Q. Cui, W. N. Lipscomb, and M. Karplus, *Angew. Chem. Int. Ed.* **42**, 1508-1511 (2003).
16. C. R. W. Guimarães, M. Udier-Blagović, I. Tubert-Brohman, and W. L. Jorgensen, *J. Chem. Theory Comput.* **1**, 617-625 (2005).
17. T. Ishida, D. G. Fedorov and K. Kitaura, *J. Phys. Chem. (B)* **110**, 1457-1463 (2006).
18. K. Kitaura, E. Ikeo, T. Asada, T. Nakano, and M. Uebayasi, *Chem. Phys. Lett.* **313**, 701-706 (1999).
19. D. G. Fedorov and K. Kitaura, *J. Chem. Phys.* **121**, 2483-2490 (2004).
20. W. D. Cornell, P. Cieplak, C. I. Bayly, I. R. Gould, K. M. Merz, Jr., D. M. Ferguson, D. C. Spellmeyer, T. Fox, J. W. Caldwell, and P. A. Kollman, *J. Am. Chem. Soc.* **117**, 5179-5197 (1995).
21. T. Ishida and S. Kato, *J. Am. Chem. Soc.* **125**, 12035-12048 (2003).
22. M. W. Schmidt, K. K. Baldridge, J. A. Boats, S. T. Elbert, M. S. Gordon, J. H. Jensen, S. Koseki, N. Matunaga, K. A. Nguyen, S. Su, T. L. Windus, M. Dupuis, and J. A. Montgomery, Jr., *J. Comput. Chem.* **14**, 1347-1363 (1993).

Quantum Fluctuations of an OCS Molecule in Superfluid Helium-4 Clusters

Shinichi Miura

Graduate School of Natural Science and Technology, Kanazawa University
Kakuma, Kanazawa 920-1192, Japan

Abstract. Using a path integral hybrid Monte Carlo method, the author examines quantum kinetic energies of an OCS molecule in helium-4 clusters as a function of the cluster-size N. The translational kinetic energy of the dopant monotonically increases with N, which is caused by the density augmentation of the helium atoms around the molecule. The effect of the Bose statistics is found to be small for the translational kinetic energy. On the other hand, the rotational kinetic energy exhibits rich size dependence, especially in a medium size regime, $10 \leq N \leq 20$. The size dependence of the rotational kinetic energy is found to be produced by the anisotropy of the density distribution around the molecule and by the superfluidity of the doped cluster.

Keywords: Superfluid, Cluster, Helium, Path integral hybrid Monte Carlo
PACS: 36.40.Mr, 61.46.+w, 67.40.Yv.

INTRODUCTION

Chemical processes in condensed helium-4 have recently been revealed to show various exotic properties due to the superfluidity of the medium [1, 2, 3, 4]. An impressive example is provided by carbonyl sulfide (OCS) molecules dissolved in helium nanodroplets [5]. The infrared spectra of the OCS molecules inside ^{4}He (boson) and ^{3}He (fermion) droplets have been measured. In the ^{4}He droplets, sharp rotational lines were observed, whereas in the ^{3}He droplets only a broad line was found. The former behavior implies that molecules rotate freely with an effective moment-of-inertia in the droplets; the latter may qualitatively be understood by the rotational diffusion in the condensed medium. Since intermolecular interaction is virtually identical for the two systems, the apparent difference observed in the spectra comes from the difference of the quantum statistics between the two systems. Then, the effective free ! rotation in the helium-4 droplet can be attributed to the superfluidity of the nanodroplet. On the other hand, Tang *et al.* [6] have experimentally determined the rotational constant of the OCS molecule B_{eff}, which is inversely proportional to the effective moment-of-inertia, in the helium-4 clusters as a function of the cluster size N in the size range $2 \leq N \leq 8$. They found the rotational constant B_{eff} monotonically decreases with increasing N; the experimental B_{eff} value reaches the nanodroplet B_{eff} value around $N = 5$, and then overshoots the nanodroplet limit, indicating the presence of a turnover in B_{eff} as a function of N with further increasing the size of the clusters.

The OCS-doped helium clusters at finite temperature have theoretically been studied by path integral Monte Carlo (PIMC) methods [7]. In the first application of the PIMC to the OCS-doped helium clusters [3], the dopant molecule was treated as a classical object fixed at the origin. Recently, the PIMC technique has been extended to handle the dopant rotation quantum-mechanically [8, 9]. Independently, based on a path integral hybrid Monte Carlo (PIHMC) method for correlated Bose fluids [10], we have developed the PIHMC to treat the system of the quantum dopant in the Bose fluids [11], and applied the method to doped helium-4 clusters OCS(^{4}He)$_N$ [11, 12, 13]. In the present study, we examine quantum fluctuations of the OCS molecule, which are measured by the translational and rotational kinetic energies, as a function of N covering a small-to-large cluster size regime: $5 \leq N \leq !64$.

METHOD

We consider the system consisting of N helium-4 atoms obeying Bose-Einstein statistics and an OCS molecule modeled as a rigid rotor. The partition function of the system Z at an inverse temperature $\beta = 1/k_B T$ is written in discretized path integral as [7, 14]

$$Z = \frac{1}{N!} \sum_{\mathscr{P}} \int \cdots \int \prod_{s=1}^{M} d\mathbf{R}^{(s)} d\Omega^{(s)} \prod_{s=1}^{M} \rho(\mathbf{R}^{(s)}, \Omega^{(s)}, \mathbf{R}^{(s+1)}, \Omega^{(s+1)}; \Delta\tau) \tag{1}$$

where $\Delta\tau = \beta/M$ is the imaginary time step and $\rho(\Delta\tau)$ is the short-time (or high-temperature) density matrix of the system. Here, $\mathbf{R}^{(s)}$ denotes the $3(N+1)$-dimensional position vector including the molecule's center-of-mass position, and $\Omega^{(s)}$ represents the molecule's orientation in the laboratory frame; the superscript s runs from 1 to M, labeling the imaginary time slice. The permutation $\mathscr{P}$ is included in the boundary condition of the path as $\mathbf{R}^{(M+1)} = \mathscr{P}\mathbf{R}^{(1)}$. In the present study, the He-He contribution in the density matrix is represented using the pair-product form of the exact two-body density matrices [7], and the He-molecule contribution is approximated using the standard factorization technique [14] accurate up to $\mathscr{O}(\Delta\tau^4)$. In the latter expression, the rotational density matrix of the molecule is included as [14]

$$\rho^{\text{rot}}(\Omega^{(s)}, \Omega^{(s+1)}; \Delta\tau) = \sum_{J=0}^{\infty} \frac{2J+1}{4\pi} P_J(\cos\gamma) e^{-\Delta\tau B J(J+1)} \tag{2}$$

where P_J is a Legendre function and γ is an angle between molecular axis of successive two time slices. The parameter B is the rotational constant of the molecule related to the moment of inertia I, $B = \hbar^2/2I$. In order to incorporate the molecular rotation in the hybrid Monte Carlo algorithm, we define the following "potential function" using the rotational density matrix: $\rho^{\text{rot}}(\Delta\tau) \equiv e^{-\Delta\tau W^{\text{rot}}(\Omega, \Omega')}$. We refer to W^{rot} as an effective potential of quantum rotation [11, 12]. In the hybrid Monte Carlo, an equation of motion is needed to generate trail configurations. We introduce a fictitious angular momentum and a fictitious moment of inertia to sample rotational fluctuations. Then, we integrate this technique into our hybrid Monte Carlo algorithms for correlated Bose fluids [10]. Further details on the method can be found elsewhere [?].

COMPUTATIONAL DETAILS

The calculated system consists of N helium-4 atoms and an OCS molecule at temperature 0.37 K. The following system-sizes are investigated: N = 5-12, 15, 20, 40, and 64. The number of discretization needed to the path integral expression is chosen to be $M = 216$ corresponding to $1/\Delta\tau = 80$ K. The rotational constant of the OCS molecule is taken from a gas-phase experimental value $B = 0.20286$ cm^{-1} [15]. The Aziz potential [16] is used as a pairwise interaction between two helium atoms. The morphed potential of Howson and Hutson [17] is adopted for the He-OCS interaction. The path integral hybrid Monte Carlo calculations have been performed for the doped Bose clusters. For comparison, the doped clusters obeying the Boltzmann statistics have been investigated, too.

RESULTS

Prior to presenting the results of the quantum kinetic energy of the OCS molecule, we first review the size dependence of the helium density distribution around the molecule and the effect of the Bose statistics on the density distribution [13]. At the smallest cluster investigated in the present study, $N = 5$, all the helium atoms are localized around the C atom where the He-OCS interaction is a minimum, forming a doughnut-type structure of the density around the molecular axis. Then, the helium atoms start to populate around other potential minima located near the O and S atoms with increasing N. At $N = 10$, the helium distribution is found in the whole region in the vicinity of the molecule. For $N > 10$, density augmentation is observed in the intermediate region between the S and C atoms in the first solvation shell, making a new peak of the density distribution. For $N > 15$, the addition of the helium atoms induces the reconfiguration of the solvation! shell; the peak near the S atom becomes lowered and shifted to the direction of the C atom along the solvation shell. At $N = 20$, the first solvation shell is nearly completed. For $N > 20$, the second solvation shell starts to be filled by the helium. In this case, the structure of the first solvation shell itself is unchanged by adding the helium atoms, although the density increase in the first solvation shell is observed for $N \leq 64$. The trend found in the size dependence of the density distribution is common to both the Bose and Boltzmann clusters. The Bose statistics makes the density distribution slightly broader. To extract the effect of the Bose statistics, we decompose the density profile on the basis of the length of the exchange cycles. At $N = 5$, the 5-body exchange cycles are observed to a large extent; in this case, the contributions from the 2- to 4-body cycles are negligible. These

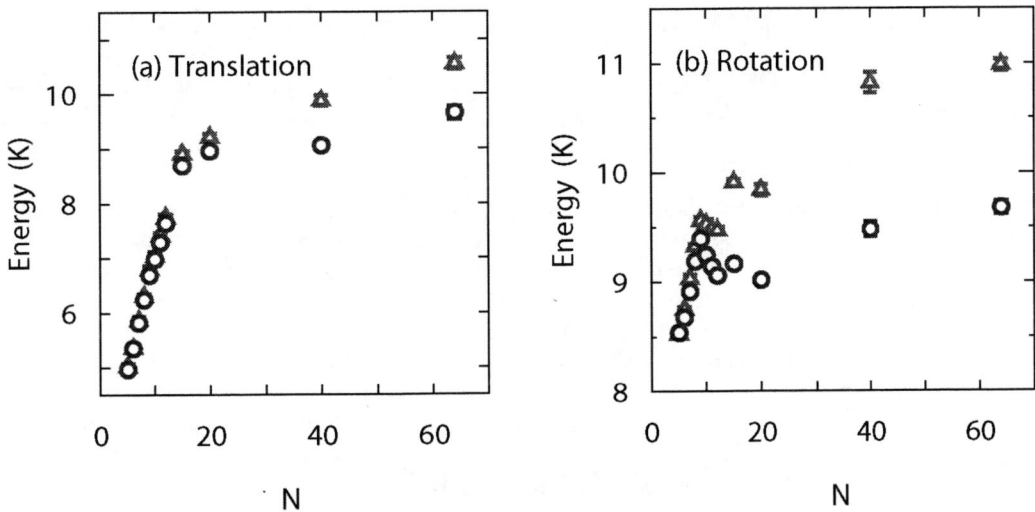

FIGURE 1. Left panel (a): The translational kinetic energy of the OCS molecule as a function of the cluster size N: N = 5-12, 15, 20, 40, and 64. Open circles represent the results of the Bose clusters and open triangles the results of the Boltzmann clusters. Right panel (b): The rotational kinetic energy of the OCS molecule as a function of the cluster size N: N = 5-12, 15, 20, 40, and 64. Open circles are for the Bose clusters and open triangles for the Boltzmann clusters. The error bar is expressed at 95 % confidence level, and is smaller than the size of the corresponding data symbol when it is not shown.

exchange cycles are located around the molecular axis in the doughnut ! structure. For $6 \leq N \leq 9$, the long-exchange cycles start! to be o bserved in the region over the C and O atoms. At $N = 10$, the bosonic exchange between the C and S regions starts to be visible, indicating the exchange cycles of the helium atoms are present to wrap the OCS molecule in the plane including the molecular axis. This type of configurations triggers the superfluidity of the doped clusters. With further increasing N, the contribution from the long exchange cycles becomes larger. For $N \geq 20$, the long exchange cycles give the dominant contribution in the density distribution.

Here, we show the size-dependence of the translational and rotational kinetic energies of the OCS molecule in Fig. 1. We first see the translational kinetic energy of the dopant molecule, K^{tra}. The kinetic energy $K^{tra}/k_B = 0.56$ K if the classical statistical mechanics is applied to the OCS molecule. Even for the smallest cluster presented ($N = 5$), the kinetic energy is found to be much larger than the classical value. The kinetic energy increase from the classical value is caused by the spatial confinement of the center-of-mass of the molecule by the surrounding helium atoms. Then, for the Boltzmann and Bose clusters, the kinetic energy becomes larger with attaching the helium atoms, which reflects the density augmentation of the helium atoms around the molecule. The effect of the Bose statistics is found to be not so large for the translational kinetic energy, although the K^{tra} for the Bose cluster is systematically smaller than the Boltzma! nn counterpart. The smaller kinetic energy found in the Bose clusters can be ascribed to the slightly broader density distribution of the helium atoms around the molecule by taking into account the Bose statistics, which gives a weaker effect of the spatial confinement.

Next, we see the rotational kinetic energy of the molecule, K^{rot}. The Boltzmann K^{rot} monotonically increases with $N(\leq 9)$ and reaches a maximum at $N = 9$. Second maximum in the size dependence is found at $N = 15$. For $N \geq 20$, the K^{rot} again increases monotonically up to $N = 64$. The non-monotonic size dependence observed in the size regime, $10 \leq N \leq 20$, can be ascribed to the anisotropic development of the solvation shell mentioned above. On the other hand, the Bose K^{rot} first shows the similar trend in the size dependence for $N \leq 9$, although, as in the K^{tra}, the Bose K^{rot} has systematically smaller value than the Boltzmann K^{rot}. In the size range $N \geq 10$, the effect of the Bose statistics becomes remarkable; the magnitude of the K^{rot} is much reduced compared with the Boltzmann counterpart. This reduction is closely related to the onset of the superfluidity of the doped cluster around $N = 10$. However, the presence of the two peaks in the size dependence is a common feature to the Bose and Boltzmann clusters, which indicates the observed trend in the K^{rot} is produced by the anisotropy of the density distribution and by the superfluidity. For $N \geq 20$, the K^{tra} slowly increases with N.

SUMMARY AND CONCLUDING REMARKS

In this paper, we have investigated the cluster-size dependence of the kinetic energy of the OCS molecule doped in the helium-4 clusters. While the translational kinetic energy is found to monotonically depend on the size N, the rotational kinetic energy shows rich size dependence produced by the anisotropy of the density distribution of the helium atoms and by the superfluidity of the doped cluster. Very recently, McKellar $et\ al.$ [18] have reported the experimental effective rotational constant B_{eff} of the OCS doped in the helium cluster up to $N = 72$. They found that the experimental B_{eff} shows unexpected oscillatory behavior in a medium-to-large size regime, $N > 20$, which does not reach the nanodroplet limit yet. Although the corresponding rotational kinetic energy may have such oscillatory size dependence, our calculations do not resolve the fine structure since only three cluster sizes, $N = 20, 40$, and 64 are investi! gated in the present study. To reproduce the broad oscillation observed in the experiment, we have to perform extensive path integral simulations covering the medium-to-large size regime. This issue will be addressed in a future study.

ACKNOWLEDGMENTS

This work was partially supported by the Grant-in-Aid for Scientific Research (No. 19550026) from the Japan Society for the Promotion of Science and by the Next Generation Super Computing Project, Nanoscience Program, MEXT, Japan.

REFERENCES

1. J. P. Toennies and D. F. Vilesov, *Annu. Rev. Phys. Chem.* **49**, 1 (1998) and references therein.
2. J. P. Toennies and D. F. Vilesov, *Angew. Chem. Int. Ed.* **43**, 2622 (2004) and references therein.
3. Y. Kwon, P. Huang, M. V. Patel, D. Blume, and K. B. Whaley, *J. Chem. Phys.* **113**, 6469 (2000) and references therein.
4. M. Barranco, R. Guardiola, S. Hernández, R. Mayol, J. Navarro, and M. Pi, *J. Low Temp. Phys.* **142**, 1 (2006) and references therein.
5. S. Grebenev, J. P. Toennies, and A. F. Vilesov, *Science* **279**, 2083 (1998).
6. J. Tang, Y. Xu, A. R. W. McKellar, and W. Jäger, *Science* **297**, 2030 (2002).
7. D. M. Ceperley, *Rev. Mod. Phys.* **67**, 279 (1995) and references therein.
8. R. E. Zillich, F. Paesani, Y. Kwon, and K. B. Whaley, *J. Chem. Phys.* **123**, 114301 (2005).
9. N. Blinov and P.-N. Roy, *J. Low Temp. Phys.* **140**, 253 (2005).
10. S. Miura and J. Tanaka, *J. Chem. Phys.* **120**, 2160 (2004).
11. S. Miura, *J. Phys.: Condens. Matter* **17**, S3259 (2005).
12. S. Miura, *J. Chem. Phys.* **126**, 114308 (2007).
13. S. Miura, *J. Chem. Phys.* **126**, 114309 (2007).
14. D. Marx and M. H. Müser *J. Phys.: Condens. Matter* **11**, R117 (1999) and references therein.
15. N. Hunt, S. C. Foster, J. W. C. Johns, and A. R. W. McKellar, *J. Mol. Spectrosc.* **111**, 42 (1985).
16. R. A. Aziz, A. R. Janzen, and M. Moldover, *Phys. Rev. Lett.* **74**, 1586 (1995).
17. J. M. M. Howson and J. M. Hutson, *J. Chem. Phys.* **115**, 5059 (2001).
18. A. R. W. McKellar, Y. Xu, and W. Jäger, *Phys. Rev. Lett.* **97**, 183401 (2006).

Spin State Dependence of Second Hyperpolarizabilities of Zethrenes

Masayoshi Nakano[*#], Ryohei Kishi[#], Akihito Takebe[#], Masahito Nate[#], Hideaki Takahashi[#], Takashi Kubo[†], Kenji Kamada[§], Koji Ohta[§], Benoît Champagne[¶], and Edith Botek[¶]

[#]Department of Materials Engineering Science, Graduate School of Engineering Science, Osaka University, Toyonaka, Osaka 560-8531, JAPAN
[†]Department of Chemistry, Graduate School of Science, Osaka University, Toyonaka, Osaka 560-0043, JAPAN
[§]Photonics Research Institute, National Institute of Advanced Industrial Science and Technology (AIST), Ikeda, Osaka 563-8577, JAPAN
[¶]Laboratoire de Chimie Théorique Appliquée Facultés Universitaires Notre-Dame de la Paix (FUNDP), rue de Bruxelles, 61, 5000 Namur, BELGIUM

Abstract. The hybrid density functional theory calculations are applied to the elucidation of the relationship between the spin state and the second hyperpolarizabilities (γ) of diradical molecules, singlet and triplet zethrenes. The longitudinal γ values of singlet zethrenes with intermediate diradical character are found to be significantly reduced in the triplet state due to the Pauli principle, while the feature is reduced when increasing the length of middle ring regions, i.e., when increasing the diradical character.

Keywords: singlet diradical, open shell, second hyperpolarizability, nonlinear optics, spin multiplicity
PACS: 42.65.An, 33.15.Kr, 31.15.Ew.

INTRODUCTION

In our previous studies [1-4], we have elucidated that singlet diradical molecules with intermediate diradical character exhibit enhanced second hyperpolarizabilities (γ) as compared to the closed-shell and pure diradical molecules. This structure-property relationship has been confirmed by using various model and real diradical compounds: H_2 dissociation [3,4], p-quinodinethane [1], twisted ethylene [1], π-conjugated molecules involving imidazole and triazole rings [5] and diphenalenyl radical compounds [6-8]. Recently, pericondensed zethrenes [9] were theoretically predicted to be singlet open-shell systems in contrast to the conventional understanding as closed-shell systems, and thus to provide much larger longitudinal γ values than the reference closed-shell molecules, tetrabenzopolyacenes [10]. The diradical characters of these zethrenes are found to be increased as increasing the length of middle ring regions, the feature of which decreases the enhancement ratio, γ(zethrene)/γ(tetrabenzopolyacene). In our previous studies, the spin state dependences of γ of diradical molecules were also examined using BI2Y [5] and diphenalenyl radicals linked by π-conjugated bridge [8], and we found that the γ values of singlet diradical systems with intermediate diradical character is significantly reduced in the triplet state, and the change is reduced as the diradical character is close to 1. Such phenomena are understood by the significant suppression of high-order polarization for the triplet spin state due to the Pauli principle. In this study, we investigate the spin state dependence of γ of zethrenes as well as its dependence on the length of middle polyacenic region by using the hybrid density functional theory (DFT) method. On the basis of the results, we show the possibility of the spin-state control of third-order NLO properties of zethrene systems.

---

[*] Corresponding author. E-mail address: mnaka@cheng.es.osaka-u.ac.jp

CP1046, Selected Papers from ICNAAM 2007 and ICCMSE 2007
edited by T. E. Simos, G. Maroulis, G. Psihoyios, and Ch. Tsitouras
© 2008 American Institute of Physics 978-0-7354-0574-5/08/$23.00

MODELS AND CALCULATIONS

Figure 1 shows the schematic structures of zethrenes, **1** ($n=4$), **2** ($n=5$) and **3** ($n=6$), where n indicates the number of benzene rings of the middle polyacenic region. The possible quinoid and diradical resonance forms suggest that these molecules could have intermediate diradical character. The molecular geometries of singlet and triplet states of these systems are optimized using the spin-unrestricted (U) B3LYP/6-31G** method under the restriction on C_{2h} symmetry. The diradical character y related to the HOMO and LUMO for singlet state is defined by the weight of the doubly-excited configuration in the multi-configurational (MC)-SCF theory and is formally expressed in the case of spin-projected UHF (PUHF) theory as

$$y = 1 - \frac{2T}{1+T^2},$$
(1)

where T is the orbital overlap between the corresponding orbital pairs [12,13] (χ_{HOMO} and η_{HOMO}) and can be represented by using the occupation numbers (n) of UHF natural orbitals (UNOs):

$$T = \frac{n_{HIOMO} - n_{LUMO}}{2}.$$
(2)

The diradical character takes values between 0 and 1, which correspond to the closed-shell and pure diradical states in the singlet state, respectively. The y values for **1-3** in singlet states calculated from the UNOs/6-31G* are given in Table 1. Molecules **1-3** in the singlet state exhibit the diradical characters y of 0.407, 0.537 and 0.628, respectively. It is noted that the increase of the length of the middle polyacenic region leads to an increase of diradical character due to the fact that the recovery of aromaticity in the linker (caused by the increase of length of linker) causes the diradical nature of both-end phenalenyl rings [13,14].

The spin-unrestricted (U) BHandHLYP/6-31G* method is employed for the calculation of longitudinal γ (γ_{xxxx}) of **1-3** in singlet and triplet states. This method is shown to be sufficient for semi-quantitative comparison of the longitudinal γ for relatively large-size hydrocarbons involving phenalenyl radicals [6-8]. We use the finite field (FF) approach [15], which consists of the fourth-order differentiation of the energy E with respect to different amplitudes of the applied external electric field. We adopt a 4-point procedure [16]:

$$\gamma = \{E(3F) - 12E(2F) + 39E(F) - 56E(0) + 39E(-F) - 12E(-2F) + E(-3F)\}/36(F)^4.$$
(3)

Here, $E(F)$ indicates the total energy in the presence of the field F applied in the longitudinal direction. We employ the values of F ranging from 0.0010 to 0.0015 a.u. in order to obtain numerically stable γ values within an error of 1 %. All calculations are performed using the Gaussian 03 program package [17].

To clarify spatial contributions of electrons to γ, the second hyperpolarizability (γ) density analysis [16] is employed. The contribution is analyzed in terms of local contributions obtained from a pair of positive and negative γ densities. The relation between γ and γ density ($\rho^{(3)}(r)$) is expressed as,

$$\gamma = -\frac{1}{3!} \int r\rho^{(3)}(r)d^3r,$$
(4)

where

$$\rho^{(3)}(r) = \frac{\partial^3 \rho(r)}{\partial F \partial F \partial F}\bigg|_{F=0}.$$
(5)

The positive and negative values of γ densities multiplied by F^3 correspond respectively to the field-induced increase and decrease in the third-order charge density, which induce the third-order dipole moment (third-order polarization) in the direction from positive to negative γ densities. Thus, the γ density map represents the relative phase and amplitude of change in the third-order charge densities between two spatial points with positive and negative values. The sign of the contribution to γ is positive when the direction from positive to negative γ density coincides with the positive direction of the coordinate system. The sign becomes negative in the opposite case. Moreover, the magnitude of the contribution associated with this pair of γ densities is proportional to the distance between them.

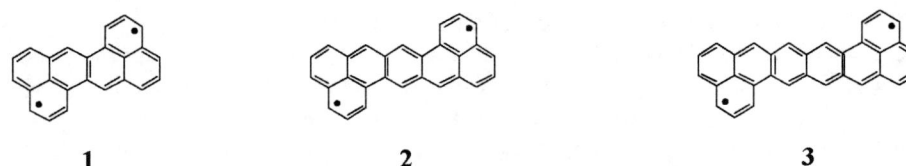

1   **2**   **3**

FIGURE 1. Schematic structures of zethrenes, **1** ($n=4$), **2** ($n=5$) and **3** ($n=6$), in singlet and triplet states. The longitudinal axis is x axis.

TABLE 1. Diradical Character y and γ Values Calculated by FF-UBHandHLYP/6-31G* Method.

	1	**2**	**3**
y (singlet)	0.407	0.537	0.628
γ (singlet) [x 10^3 a.u.]	375	863	1442
γ (triplet) [x 10^3 a.u.]	116	333	872

RESULTS AND DISCUSSION

The calculated longitudinal γ values of our model molecules are given in Table 1. For the triplet state, the UBHandHLYP method is also expected to give a reliable γ value since the spin contamination is small ($<S^2>$ = 2.25 (**1**), 2.29 (**2**) and 2.34 (**3**)). As shown in our previous study [10], the γ values of singlet zethrenes are significantly enhanced compared to those of similar-size closed-shell molecules, tetrabenzopolyacenes, whereas the increase ratio of γ(singlet zethrene)/γ(tetrabenzopolyacene) is found to be reduced when the length of middle polyacenic region increases. This is due to the increase of diradical character in the case of zethrenes with longer middle region. From Table 1, the spin state dependence of γ is also substantial. The γ values for the singlet states are larger than those of triplet states, while the increase ratio, γ(singlet)/γ(triplet), becomes small as increasing the length of middle polyacenic region: 3.2 ($n=4$), 2.6 ($n=5$) and 1.7 ($n=6$). This significant difference of γ value between the singlet and triplet states is predicted to be caused by the difference of the field-induced electron fluctuation effect between the singlet and triplet states: the large third-order polarization in the singlet state with intermediate diradical character is caused by the intermediate overlap between a pair of radical orbitals [8], while in the triplet state (the highest spin state for diradical system), such polarization is significantly suppressed because of the non-overlap between a pair of radical orbitals due to the Pauli principle. Because in the region with large diradical character for the singlet state, the overlap between diradical distributions becomes small for both spin states, such difference of γ between the singlet and triplet states tend to be reduced. This is exemplified by the reduction of increase ratio, γ(singlet)/γ(triplet), as increasing n value.

We next examine the comparison of γ density distributions for these molecules (Figure 2). As shown in our previous paper, the main contribution to γ is found to come from π-electrons, which give positive contribution in

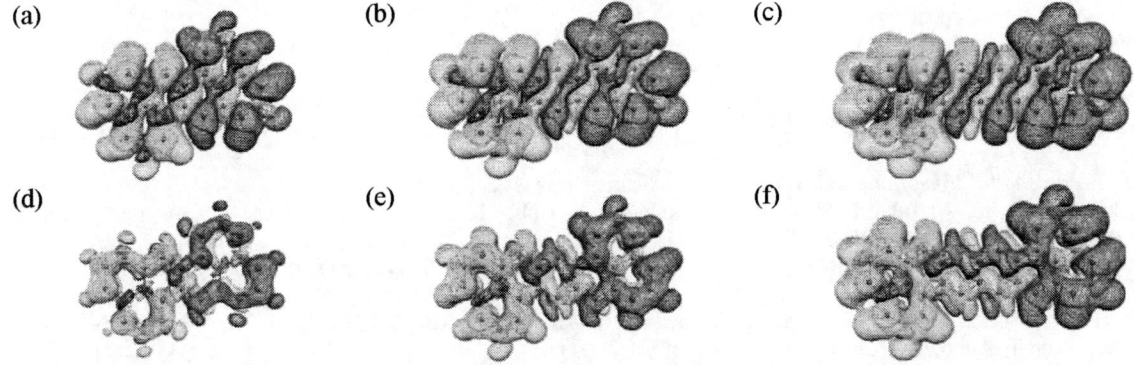

FIGURE 2. γ density distributions for total electron contributions in the singlet [(a)-(c)] and triplet [(d)-(f)] zethrenes. The yellow and blue meshes represent positive and negative densities with iso-surfaces with ±100 a.u., respectively.

contrast to σ-electrons with negative one. The main contributions to γ of singlet zethrenes (a)-(c) turn out to be caused by the extended positive and negative γ densities well separated on the left and right phenalenyl ring regions, respectively, though in the middle regions the γ densities present sign alternation. For the triplet zethrenes (d)-(f), the amplitudes of γ densities are significantly reduced as compared to those of the singlet zethrenes though the difference becomes small in the long middle polyacenic region. Moreover, for longer chains, the middle region of the triplet zethrene displays an *up-down polarization* of the γ densities, which alternate in sign.

All these features well support the relative features of γ values between singlet and triplet zethrenes.

CONCLUDING REMARKS

We have investigated the spin state dependences of longitudinal γ for zethrenes with different lengths of midlle polyacenic regions. It is found that the γ values of triplet states of zethrenes, corresponding to the highest spin states for diradical systems, are significantly reduced as compared to those of singlet states. This is predicted to be caused by the reduction of field-induced electron polarization due to the Pauli principle. It is also found that the ratio, γ(singlet)/ γ(triplet), decreases when increasing the length of middle polyacenic region, which is caused by the increase of diradical character when elongating the middle polyacenic region. Namely, the field-induced electron polarization in the singlet state gets smaller for large diradical character so that the radical orbital pattern of the triplet and singlet states becomes similar as increasing the length of middle polyacenic region. The structure-property relationship obtained in this study will contribute to the development of spin state control of third-order NLO properties and sensor devices of spin states.

ACKNOWLEDGMENTS

This work is supported by Grant-in-Aid for Scientific Research (No. 18350007) from Japan Society for the Promotion of Science (JSPS), and Grant-in-Aid for Scientific Research on Priority Areas (No. 18066010) from the Ministry of Education, Science, Sports and Culture of Japan. E.B. thanks the IUAP program N° P6-27 for her postdoctoral grant. B.C. thanks the Belgian National Fund for Scientific Research for his Research Director position.

REFERENCES

1. M. Nakano, R. Kishi, T. Nitta, T. Kubo, K. Nakasuji, K. Kamada, K. Ohta, B. Champagne, E. Botek and K. Yamaguchi, *J. Phys. Chem. A* **109**, 885 (2005).
2. B. Champagne, E. Botek, M. Nakano, T. Nitta and K. Yamaguchi, *J. Chem. Phys.* **122**, 114315-1 (2005).
3. M. Nakano, R. Kishi, S. Ohta, A. Takebe, H. Takahashi, S. Furukawa, T. Kubo, Y. Morita, K. Nakasuji, K. Yamaguchi, K. Kamada, K. Ohta, B. Champagne and E. Botek, *J. Chem. Phys.* **125**, 74113 (2006).
4. M.Nakano and K. Yamaguchi, *Phys. Rev. A* **50**, 2989 (1994).
5. M. Nakano, R. Kishi, N. Nakagawa, S. Ohta, H. Takahashi and S. Furukawa, K. Kamada, K. Ohta, B. Champagne, E. Botek, S. Yamada and K. Yamaguchi, *J. Phys. Chem. A* **110**, 4238 (2006).
6. M. Nakano, T. Kubo, K. Kamada, K. Ohta, R. Kishi, S. Ohta, N. Nakagawa, H. Takahashi, S. Furukawa, Y. Morita, K. Nakasuji and K. Yamaguchi, *Chem. Phys. Lett.* **418**, 142 (2005).
7. S. Ohta, M. Nakano, T. Kubo, K. Kamada, K. Ohta, R. Kishi, N. Nakagawa, B. Champagne, E. Botek, S. Umezaki, A. Takebe, H. Takahashi, S. Furukawa, Y. Morita, K. Nakasuji and K. Yamaguchi, *Chem. Phys. Lett.* **420**, 432 (2006).
8. S. Ohta, M. Nakano, T. Kubo, K. Kamada, K. Ohta, R. Kishi, N. Nakagawa, B. Champagne, E. Botek, A. Takebe, S. Umezaki, M. Nate, H. Takahashi, S. Furukawa, Y. Morita, K. Nakasuji and K. Yamaguchi, *J. Phys. Chem. A* **111**, 3633 (2007).
9. E. Clar, *Polycyclic Hydrocarbons*, vol. 1, London: Academic Press, 1964, pp. 32.
10. M. Nakano, R. Kishi, A. Takebe, M. Nate, H. Takahashi, T. Kubo, K. Kamada, K. Ohta, B. Champagne and E. Botek, *Computing Letters*, in printing, (2007).
11. K. Yamaguchi, in: R. Carbo, M. Klobukowski (Eds.), *Self-Consistent Field: Theory and Applications*, Amsterdam: Elsevier, 1990, pp. 727.
12. S. Yamanaka, M. Okumura, M. Nakano and K. Yamaguchi, *J. Mol. Struct.* **310**, 205 (1994).
13. K. Nakasuji and T. Kubo, *Bull. Chem. Soc. Jpn.* **77**, 1791 (2004).
14. K. Nakasuji, K. Yoshida and I. Murata, *Chem. Lett.* 969 (1982).
15. H.D. Cohen and C.C.J. Roothaan, *J. Chem. Phys.* **43**, S34 (1965).
16. M. Nakano, I. Shigemoto, S. Yamada and K. Yamaguchi, *J. Chem. Phys.* **103**, 4175 (1995).
17. M. J. Frisch et al., *GAUSSIAN 03, Revision B.04*, Gaussian Inc., Pittsburgh, PA, (2003).

Multireference Density Functional Study of Atomic and Molecular Magnetic Systems

Kazuto Nakata [1], Shusuke Yamanaka [2,3], Takeshi Ukai [1],
Toshikazu Takada [3,4], Kizashi Yamaguchi [2,3]

1) Department of Chemistry, Graduate School of Science, Osaka University, Toyonaka, 560-0043, Japan.
2) Center for Quantum Science and Technology under Extreme Conditions, Osaka University, Toyonaka, 560-8531, Japan
3) CREST Project, Japan Science and Technology Agency, 4-1-8 Honcho, Kawaguchi City, Saitama 332-0012, Japan
4) RIKEN, Next-Generation Supercomputer R&D center, Chiyodaku, Tokyo, 100-0005, Japan

Abstract. We present the complete-active-space density functional theory (CAS-DFT) approach for degenerate and nearly degenerate systems. In this method, we solve the effective CAS-DFT equation in order that the Euler equation of CAS-DFT is satisfied. The simple applications for atomic and molecular magnetic systems such as atomic multiplets, organic radicals, and Cu(II) acetate are presented. The results are discussed in relation to the theory of magnetism.

Keywords: Multireference effects; Density functional theory; Molecular Magnetism.
PACS: 31.15.Ar; 31.15.Ew; 36.40.Cg

INTRODUCTION

The strategy to treat the electron correlation effects in the Kohn-Sham density functional theory (KS-DFT) is usually based on the semi-local approximation of exchange-correlation (XC) effects within the noninteracting picture, assuming homogeneous or nearly homogeneous electron gas system [1]. This strategy leads to the great success in the field of molecular science for the reason that it enables us to treat the large systems [2]. However it is noteworthy that the XC functional has been considerably improved from the time that the KS-DFT is introduced in the computational chemistry: starting the local density approximation (LDA), the dependence on the gradients of density, that on kinetic energy density, and that on occupied orbitals are considered, leading to generalized gradient approximation (GGA), meta-GGA, and hybrid DFT, respectively. Adding the optimized effective potential (OEP) method that depends on not only on the occupied, but also on unoccupied orbitals [3], Perdew and his co-workers arrange this sequence of XC functionals and call it "Jacob's ladder of DFT" [4]. A standard approach commonly employed in the computational chemistry is the hybrid DFT approach, in which the exact Hartree-Fock (HF) exchange is mixed with the pure DFT XC functional. The reason to require the hybrid-DFT might be due to the fact that the exchange term of atomic and molecular systems is not described well if we employ the pure semi-local approximation such as LDA, GGA, and meta-GGA. In these approximations, the exchange hole is assumed to be spherical, but it is known that this approximation deviates considerably from the exact situation for molecular systems. The exact exchange hole reflects the structure of the molecular system we consider. Now note that a similar argument is possible for the correlation term. The usual semi-approximated correlation functional does not reflect the molecular structure on its functional form. By contrast, correlation functionals at the top of the Jacob's ladder of DFT cover the structure dependent correlation effects via the dependency on occupied-and-unoccupied orbitals. This type of DFT corresponds to the many-body perturbation theory based on the single-determinant theory such as Møller-Plesset (MP) perturbation theory (PT) and coupled-cluster (CC) theory [5,6]. The correlation functional at this rung on the Jacob's ladder is expected to be useful for the phenomena, for which virtual excitation processes of electrons among orbitals play an important role. Indeed, the weak-molecular interactions based on the dispersion effects among closed-shell molecules can be reproduced well by OEP based on the PT [5].

CP1046, *Selected Papers from ICNAAM 2007 and ICCMSE 2007*
edited by T. E. Simos, G. Maroulis, G. Psihoyios, and Ch. Tsitouras
© 2008 American Institute of Physics 978-0-7354-0574-5/08/$23.00

For molecular magnetism, however, a different approach is necessary to provide an accurate description of electronic structure, because another type of correlations plays an important role: remember that for many effects to determine the magnetic structure of compounds, the multireference wavefunction involved the low-lying excitations, i.e. CAS wavefunction, is assumed [7,8]. This type of correlations is considerably different from that covered by semi-local correlation functional and that covered by the orbital-dependent PT. In order to provide an overview of such a situation, we previously classified the electron correlations into three types, structure-independent, orbital-dependent, and resonating [9]. These correlations overlap each other as shown in figure 1 (A). However, it is difficult to cover the specific type of correlation effects by an ab initio method that covers other type. The predominant type of correlations depends on the character of the system: for metal states of solids, the structure-free type of correlations covered by LDA is important while the orbital-dependent type is essential to study van der Waals systems. On the other hand, the important type to determine magnetism of materials is the resonating type.

In the contemporary ab initio computational chemistry, the resonating effects are covered by CAS wavefunction. In the context of DFT, the XC functional to cover this type of effects is unknown (and may be non-existing for many cases). Thus, the prescription to treat resonating effect is to replace the single-determinant picture of KS-DFT by the multireference picture of MR-DFT [9-20].

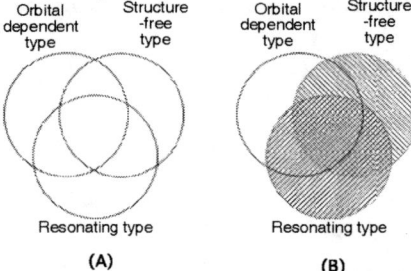

FIGURE 1. (A) Three types of electron correlations. (B) Electron correlations covered by CAS-DFT.

Complete-active-space density functional theory

The fundamental equation of CAS-DFT is that of the Hohenberg-Kohn-Levy's variational principle.

$$E = Min_\rho \left[F^{CAS}[\rho(\mathbf{r})] + \int d\mathbf{r}\rho(\mathbf{r})V_{ext}(\mathbf{r}) + E_{RC}[\rho(\mathbf{r})] \right] \tag{1}$$

Here, $V_{ext}(\mathbf{r})$ is the external potential of the system and $F^{CAS}[\rho(\mathbf{r})]$ is the modified universal functional given by,

$$F^{CAS}[\rho(\mathbf{r})] = Min_\rho^{CAS} \langle \Psi^{CAS} | \hat{T} + \hat{V}_{ee} | \Psi^{CAS} \rangle. \tag{2}$$

Here the search region is limited to be the space expanded by specific type of CAS wavefunctions. This is different from the corresponding "noninteracting kinetic functional" of KS-DFT, $T_s[\rho(\mathbf{r})] = Min_\rho \langle \Phi | \hat{T} | \Phi \rangle$. Incorporating the electron-repulsion operator into the modified universal interaction as in eq. (2), we intend to cover "resonating type" of correlations in figure 1 (B) by this functional. The remaining correlation effects are included in the residual correlation (RC) term in eq. (1), which is denoted as $E_{RC}[\rho(\mathbf{r})]$. In order to satisfy the Euler-equation,

$$\delta F^{CAS}[\rho(\mathbf{r})]/\delta\rho(\mathbf{r}) = -V_{ext}(\mathbf{r}) - \delta E_{RC}[\rho(\mathbf{r})]/\delta\rho(\mathbf{r}) + \mu. \tag{3}$$

we solve the CASCI-DFT equation given by

$$\left(\hat{T} + \hat{V}_{ee} + \hat{V}_{ext} + \hat{V}_{RC} \right) |\Psi\rangle = E_{MR-DFT} |\Psi\rangle. \tag{4}$$

where $\langle \mathbf{r} | \hat{V}_{RC} | \mathbf{r} \rangle \equiv \delta E_{RC}[\rho(\mathbf{r})]/\delta\rho(\mathbf{r})$ is the RC potential. In this formulation, the modified universal functional is defined with specifying a wavefunction expansion in eq. (2). But, note that another partition scheme is possible:

dividing the Coulomb repulsion operator into short and long range parts and incorporating the later into the modified universal functional, Savin and his coworkers have derived the Coulomb-driven MR-DFT [11,17]. This scheme allows us to combine any wavefunction method with DFT, such as MP2-DFT and CC-DFT [18]. We intend to cover all of exchange term by the wavefunction part [15], because the exact exchange plays an important role in magnetism of materials, so that we employ the wavefunction-driven MR-DFT as described above. In the context of DFT ladder, this implies that the explicit dependence on truncated CI coefficients is used in DFT. The extension to CASSCF-DFT approach is straightforward [16]. Further, it is possible to exploit both CI coefficients and occupied-and-unoccupied orbitals as the explicit variables in the context of MR-DFT, leading to the MR-DFT with orbital dependent corrections [9] that mimics the MR-MP approach.

In this study, however, for simplicity, we employ the usual Lee-Yang-Parr functional with the prescription proposed by MSS and GC [12], with which the double-counting part of the semi-local correlation functional for CAS-DFT is filtered out. Further, another auxiliary variable for RC functional [13,14,19,20] that corresponds to spin-polarizability in the KS-DFT is omitted. The electron correlations covered by this approach corresponds to the shaded region in figure 1 (B). In this study, we expect that the resonating type and structure-dependent type of correlations are predominantly important to determine the electronic structures of atoms and molecules.

Results and Discussions

Spectrum of V (III) ion with ligand field

The ground-state of the Vanadium ion (V(III)) has ^3F symmetry [21], which can not be expressed by a single determinant, as illustrated as figure 2(A). Now we apply the CASSCF-DFT approach with employing two electrons and five d orbitals as an active space and Watchers+f basis [22]. The resulting spectra of V(III) are listed in table 1. The excitation energies are slightly larger than the experimental one, of which situation is similar to other CAS plus correlation corrections. The order of spectra is similar to that of experimental one, and differences between calculated and experimental values are small for up to the third excited state.

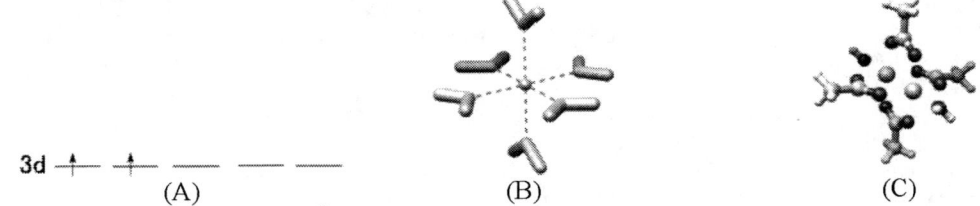

FIGURE 2. (A) A configuration of ^3F state of V(III). Geometries of V(III) with six H_2O. (B) and Cu(II) acetate complex(C).

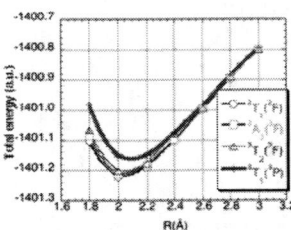

FIGURE 3. Low-lying spectra of V(III) with six waters (illustrated in figure 2(B)).
Further we examine low-lying states of V(III) ion with six waters, as illustrated in figure 2(B). Spectra for various interatomic distances between H_2O and V(III) are plotted in figure 3. These results are consistent with those of ligand-field theory [23]. The results of V(III) and its complex coordinated with waters imply that the CASSCF-DFT can be used as a reliable computational approach for degenerate system of mono-transition metal (TM) complex.

TABLE 1. Low-lying spectra of V(III) calculated by CASSCF-DFT

	$^3F \rightarrow ^1D$	$^3F \rightarrow ^3P$	$^3F \rightarrow ^1G$	$^3F \rightarrow ^1S$
CASSCF-DFT	1.682	2.017	2.631	6.347
Experimental value	1.359	1.645	2.280	—

Now we examine Copper (II) acetate ($Cu_2(OAc)_4(H_2O)_2$) of which structure is illustrated in figure 2(C), as an example of dinuclear-TM complexes. This compound is known to exhibit strong negative J value ($-148cm^{-1}$) [24] and origin of such antiferromagnetic (AFM) interactions is a controversy issue. We apply CASCI[2,2]-DFT and CASCI[6,6]-DFT using the natural orbitals of UHF solutions to estimate J value of this species. Together with the experimental and UHF results, calculated values are listed in table 2. It is obvious that J value of CASCI[6,6]-DFT is similar that of experiment, while that of CASCI[2,2]-DFT is considerably smaller than the experimental value. The dominant contribution from CAS [2,2] to magnetic interaction is direct exchange between two Cu(II) whereas those from CAS[6,6] are superexchange interactions via acetate ligands. Thus we can conclude that the origin of the strong AFM interaction is the superexchange interactions via acetate ligands.

Also this result implies that CAS-DFT with an appropriate active space is useful to investigate magnetic interactions of dinuclear TM complex.

TABLE 2. Magnetic interactions of $Cu_2(OAc)_4(H_2O)_2$ calculated by CASCI-DFT, UHF, together with experimental one

	CASCI[2,2]-DFT	CASCI[6,6]-DFT	UHF	Experimental value
J / cm^{-1}	-4.81	-160.4	-20.4	-148

ACKNOWLEDGMENTS

This research is supported by CREST project by Japan Science and Technology Agency (JST).

REFERENCES

1. W. Kohn, and L. J. Sham, Phys. Rev. **140**, A1131 (1964).
2. W. Koch and and M. C. Holthausen, *A Chemist's Guide to Density Functional Theory*, Wiley-VCH, Weinheim, 2000.
3. A. Görling, and M. Levy, Phys Rev **A 50**, 196 (1994)
4. J. P. Perdew and K. Schmidt, in *Density Functional Theory and Its applications to materials,* edited by V. V. Doren, K. V. Alsenoy, and P. Geerlings, AIP Conference Proceedings 620, American Institute of Physics, Melville, NY, 2001.
5. E. Engel, and R. Dreizler, J. Comp. Chem., **20**, 31 (1999)
6. R. J. Bartlett, V. F. Lotrich, and I. V. Schweigert, J. Chem. Phys. **123**, 062205 (2005).
7. K. Yoshida, Theory of Magnetism, Springer, Berlin, 1996.
8. K. Yamaguchi, In *MolecularMagnetism*, ed.K. Ito and M.Kinoshita, Gordon and Breach, Amsterdam,2000.
9. S. Yamanaka, K. Nakata, T. Ukai, T. Takada, and K. Yamaguchi, Int. J. Quantum Chem. **106**, 3312 (2006).
10. A. Savin Int. J. Quantum Chem., **34**, 59 (1988).
11. A. Savin, and H-J. Flad, Int. J. Quantum Chem., **56**, 327 (1995).
12. B. Miehlich, H. Stoll, and A. Savin, Mol. Phys., **91**, 527 (1997); J. Gräfenstein, and D. Cremer, Chem. Phys. Lett. **316**, 569 (2000).
13. S. Gusarov, P-A. Malmqvist, R. Lindh, B. O. Roos, B. O. Theor Chem Acc. **112** , 84, (2004).
14..R. Takeda, S. Yamanaka, and K. Yamaguchi, Chem. Phys. Lett. **366**, 321 (2002); S. Yamanaka, T. Ohsaku, D. Yamaki, and K. Yamaguchi, Int. J. Quantum Chem. **91**, 376 (2003); R. Takeda, S. Yamanaka, and K. Yamaguchi, Int. J. Quantum Chem. **96**, 463 (2004).
15. S. Yamanaka, K. Nakata, T. Takada, K. Kusakabe, J. M. Ugalde, and K. Yamaguchi, Chem. Lett. **35**, 242 (2006).
16..K. Nakata, T. Ukai, S. Yamanaka, T. Takada, and K. Yamaguchi, Int. J. Quantum Chem. **106**, 3325 (2006); T. Ukai, K. Nakata, S. Yamanaka, T. Kubo, Y. Morita, T. Takada, and K. Yamaguchi, Polyhedron, **26**, 2313 (2007).
17. P. Gori-Giorgi, A. Savin, Phys Rev A **73**, 032506 (2006) ; J. Toulouse, F. Colonna, A. Savin, Phys Rev A, **70**, 062505 (2004).
18. I. C. Gerber and J. G.. Angyan, J. Chem. Phys. **126**, 044103 (2007).
19. F Moscardò and E. San-Fabián, Phys Rev **A44**, 44, 1549 (1991).
20. A. J. Pérez-Jiménez and J. M. Pérez-Jord, Phys. Rev. **A 75,** 012503 (2007).
21. C. E. Moore, *Atomic Energy Levels*, Natl. Bur. Stand , U.S. Circ. No. 476,U.S. GPO, Washington, D.C., 1971.
22. J. H. Wachters, J Chem Phys **52**; 1033 (1970); C. W. Bauschlicher, C. W., Jr., S.R. Langhoff, L. A. Barnes, J Chem Phys **91**, 2399 (1989).
23. S. Sugano, Y. Tanabe, and H. Kamimura, *Multiplets of transition-metal ions in crystals,* Pure and applied physics : a series of monographs and textbooks v. 33 Academic Press, New York, 1970.
24. R.L. Carlin, *Magnetochemistry*, Springer, Berlin Heidelberg, 1986;P. De Meester, S.R. Fletcher, A.C. Skapski, J. Chem. Soc., Dalton Trans. 2575, (1973).

Modeling of Hydrogen Storage Materials: A Reactive Force Field for NaH

Ojwang' J.G.O.*, Rutger van Santen*, Gert Jan Kramer*, Adri C. T. van Duin† and William A. Goddard III†

*Schuit Institute of Catalysis, Eindhoven University of Technology, Postbus 513, 5600 MB, Den Dolech 2, Eindhoven, The Netherlands.
†Material Research Center, California Institute of Technology(Caltech) , 1200 East California Boulevard Pasadena, California 91125, U.S.A.

Abstract. Parameterization of a reactive force field for NaH is done using *ab initio* derived data. The parameterized force field(ReaxFF$_{NaH}$) is used to study the dynamics governing hydrogen desorption in NaH. During the abstraction process of surface molecular hydrogen charge transfer is found to be well described by the parameterized force field. To gain more insight into the mechanism governing structural transformation of NaH during thermal decomposition a heating run in a molecular dynamics simulation is done. The result shows that a clear signature of hydrogen desorption is the fall in potential energy surface during heating.

Keywords: hydrogen storage, reactive force field, abstraction, desorption, cluster, metastable state
PACS: 64.30.+t, 31.15.Ew, 36.20.Ey

INTRODUCTION

Bogdanovic [1] pioneered the interest in the complex metal alanates by establishing that the thermal decomposition process of NaAlH$_4$ could be reversed by doping with metal catalysts of which, titanium is the most thoroughly investigated. However, there are still unresolved problems in understanding of H$_2$ dissociation process in the system. Among these include the role of titanium (is it a dopant or a catalyst?) and long range transport mechanism of Al during the dissociation process.

Briefly, the conventional (de)sorption steps occurs as follows:

$$NaAlH_4 \quad \leftrightarrow \quad \frac{1}{3}Na_3AlH_6 + \frac{2}{3}Al + H_2 \quad (3.7wt\%) \tag{1}$$

$$Na_3AlH_6 \quad \leftrightarrow \quad 3NaH + Al + \frac{3}{2}H_2 \quad (1.9wt\%) \tag{2}$$

We are developing a reactive force field(ReaxFF$_{NaAlH_4}$) to study the structural and dynamical details of hydrogen (de)sorption processes in NaAlH$_4$ system. We aim to elucidate on the long range transport mechanisms of Al atoms during (de)sorption process and the dynamics governing hydrogen desorption. Towards this end we are parameterizing a force field, ReaxFF[2], to simulate large clusters containing NaH, Na$_3$AlH$_6$, NaAlH$_4$ and Al phases plus catalysts atoms. ReaxFF has already been shown to be able to accurately predict the dynamical and reactive processes in hydrocarbons[2], silicon/silicon oxides[3], aluminum/aluminum oxides[4] and nitramines[5].

As a run-up towards parameterizing ReaxFF$_{NaAlH_4}$, we have already sufficiently parameterized ReaxFF$_{NaH}$, which is a reactive force field for sodium and sodium hydride, to adequately describe H$_2$ desorption process in NaH. In this work the details of the parameterizations of ReaxFF$_{NaH}$, the diffusion mechanism of hydrogen atoms and hydrogen molecules in NaH and abstraction process of surface molecular H$_2$ in NaH cluster are examined. It will be shown that a clear signature for H$_2$ dissociation is the fall in potential energy surface during heating process in a molecular dynamics simulation run.

CP1046, *Selected Papers from ICNAAM 2007 and ICCMSE 2007*
edited by T. E. Simos, G. Maroulis, G. Psihoyios, and Ch. Tsitouras
© 2008 American Institute of Physics 978-0-7354-0574-5/08/$23.00

REACTIVE FORCE FIELD FOR SODIUM AND SODIUM HYDRIDE

Parametrization of ReaxFF$_{NaH}$ has been done in line with those used to develop ReaxFF$_{MgH}$[6]. Initially developed for hydrocarbons[5], ReaxFF has been successfully applied to study Si/SiO$_2$ interfaces[3], MgH$_2$ systems and Al/α-Al$_2$O$_3$ systems[4]. A key feature in ReaxFF is using the bond-order formalism that allows for bond breaking and formation as per Tersoff[7], Brenner[8] and environment dependent interatomic potential(EDIP)[9] approaches. Charge calculation are fitted using electronegativity equalization method(EEM)[10], which allows for polarizability and geometry dependent charge distribution. ReaxFF calculates non-bonded (van der Waals and Coulomb) interactions between all atoms (including 1-2 and 1-4 interactions) making it suitable for systems in which there are covalent and ionic interactions. It is this last feature, coupled with the ability to create and annihilate bonds, that makes ReaxFF attractive for modeling NaAlH$_4$ in which there is an interplay of both polar and covalent interactions.

The quantum values used in the fitting were obtained from VASP[11], using projector augmented[12] plane-wave method. The calculations used the generalized gradient approximation of Perdew and Wang[13, 14, 15]. The partial charges used to optimize the EEM parameters were obtained by performing a Mulliken charge distribution analysis using CRYSTAL06[16], which implements a periodic localized basis set (LCAO) approach.

The total energy expression in ReaxFF is partitioned into several contributions as follows [3]:

$$E_{sys} = E_{bond} + E_{over} + E_{under} + E_{val} + E_{NaH} + E_{vdWaals} + E_{Coulomb} \qquad (3)$$

Parameterizations of the energy expressions were done by fitting into the training set the *ab initio* derived equations of state (EoS) of pure Na and NaH condensed phases, reaction energies and bond dissociation profiles on small finite clusters. Phase transformations/crystal modifications in both Na and NaH systems during desorption process was accounted for by adding the high pressure phases of Na and NaH, in addition to the groundstate phases, to the quantum calculations. In the case of Na we considered four phases: bcc-Na(8-coordinate), sc-Na(6-coordinate), fcc-Na(12-coordinate) and hcp-Na(12-coordinate). For NaH, the high pressure (CsCl-type) and NaCl-type phases were considered.

Bond dissociation

Additional tests for the reactive potential were conducted by taking into account density functional theory(DFT) values of bond dissociation profiles of small NaH clusters. Figure 1(a) shows the bond dissociation curve of NaH. ReaxFF gives an equilibrium bondlength of 1.911 Å which is in excellent agreement with DFT value of 1.929 Å. Figure 1(b) shows the NaN-NaH bond dissociation curve in which Na$_2$H$_2$ fragments into two NaH. DFT gives an equilibrium NaH-NaH bondlength of 2.076Å while ReaxFF gives 2.108 Å. For each of these cases both DFT singlet and triplet states were considered. The bond parameters were then optimized to the lowest energy point along the dissociation profile.

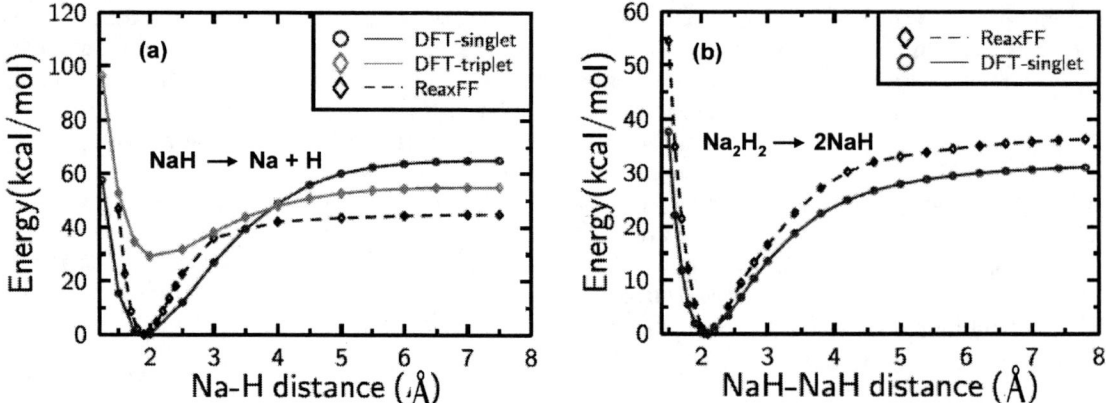

FIGURE 1. Bond dissociation curves of small clusters of NaH as calculated by DFT and ReaxFF. ReaxFF correctly captures both the short and long range behavior of the Na-H bond.

Equations of state

For both Na and NaH, the ability of ReaxFF to capture the relative phase stability was tested against a number of Na and NaH crystal modifications.

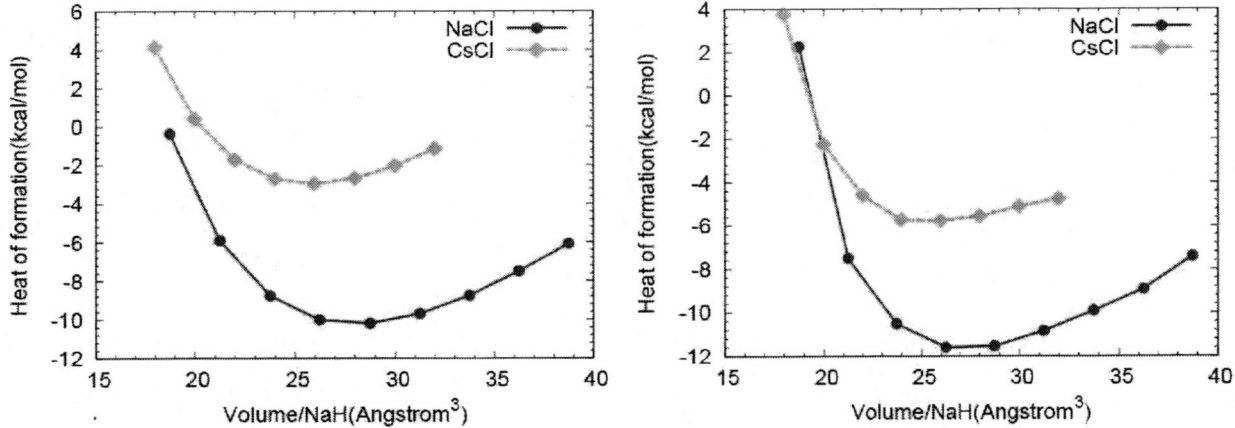

FIGURE 2. Equations of state of NaH as computed from DFT and ReaxFF respectively.

For each an every phase of Na(fcc, hcp, sc and bcc) and NaH(NaCl and CsCl types) considered in this work, the quantum energies were computed for abroad range of volume expansion and compression. The phase stabilities of ReaxFF values were compared against DFT values, Fig. 2 and table 1. There is an excellent match between DFT and ReaxFF values.

TABLE 1. Na metal and NaH phase stabilities(kcal/mol) relative to the bcc and NaCl-type phase respectively.

Na metal phase stabilities		
Phase	ΔE_{DFT}	ΔE_{ReaxFF}
HCP	0.54	0.58
FCC	0.24	0.34
SC	1.67	3.78
NaH phase stability		
CsCl-type	7.24	5.83

ABSTRACTION OF HYDROGEN

In order to get a better insight of crystal transformation during the desorption of hydrogen, we simulated successive abstraction of surface molecular hydrogen from $Na_{48}H_{48}$ cluster. The abstraction process is given by:

$$Na_{48}H_{n+2} \rightarrow Na_{48}H_n + H_2 \qquad (4)$$

where n = 46 to 0.

In the abstraction process, clusters were first minimized to find the nearest metastable conformation and then equilibrated at 300 K for 50000 steps(timestep was set at 0.25 fs). The clusters were then annealed to 0 K before abstraction of H_2 molecule as in the starting configuration of each simulation was the last configuration of the previous simulation's annealed run less two H atoms. This was done iteratively until all the H_2 atoms were removed.

There are two stages shown in Fig. 3(a), which shows that there is a non-linear trend in particle stability with respect to molecular hydrogen abstraction. During the abstraction process, the surface hydrogen atoms are removed first. As a result of this, some of the bulk Na atoms comes to the surface to replace the depleted hydrogen atoms. In the starting

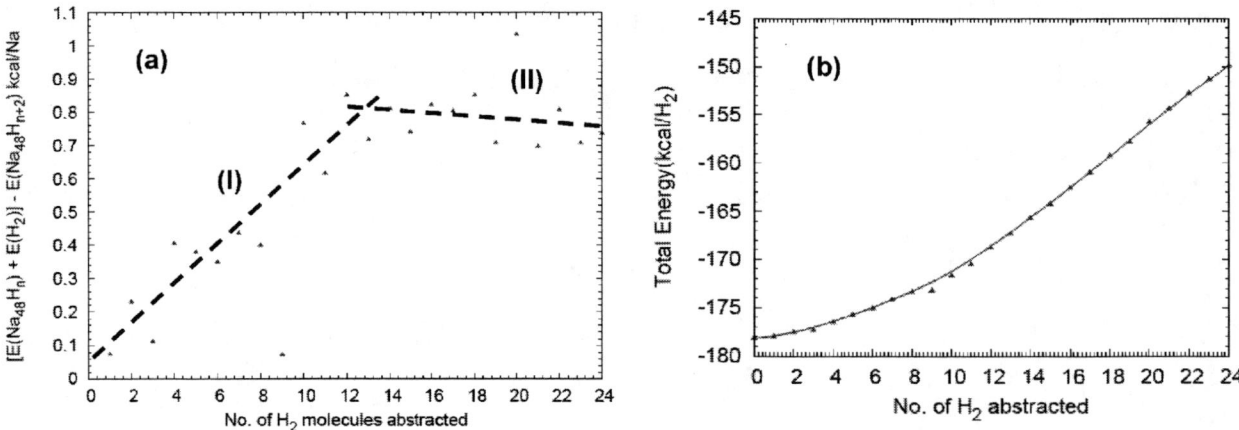

FIGURE 3. (a) Abstraction energy, $\Delta E = [E(Na_{48}H_n) + E(H_2)] - E(Na_{48}H_n + 2)kcal/Na$ and (b) Total energy as a function of number of H_2 molecules abstracted from the system.

structure, the surface hydrogen atoms mostly occupy the less stable two fold (bridge) sites. However, as more and more surface molecular hydrogen are abstracted the remaining hydrogen atoms adopts a subsurface conformation and they occupy three and four fold sites. With increasing depletion of hydrogen atoms, the sodium atoms in the bulk replaces the depleted hydrogen atoms. Therefore, the hydrogen atoms are more strongly bound in the system. This explains the trend in part (I) of Fig. 3. What is interesting about Fig. 3(a) is the change in the graph, (II), after slightly more than a half of the hydrogen atoms have been abstracted from the system. The reason for this change is that metallization dominates over ionization. We can think of this stage as a situation whereby hydrogen atoms are dissolved in Na metal matrix, with the metal retaining its metallic nature. Figure 3(b) shows the total energy as a function of the number of H_2 molecules. The graph shows that with increasing abstraction of H_2 molecules the system becomes less stable. The structure at the end, $48Na + \frac{48}{2}H_2$, is less stable than the starting structure, $Na_{48}H_{48}$. This is expected since the heat of formation of NaH is negative indicating that it is more stable with respect to its constituent elements. The scattering in Fig. 3(a) is related to the rearrangement and cluster fragmentation during the abstraction process. In other words, abstraction destabilizes the system.

MOLECULAR DYNAMICS SIMULATION

One of the goals of ReaxFF$_{NaH}$ is to investigate the thermally induced desorption of hydrogen molecules in NaH. To do this, we performed a NVT simulation on a small cluster of NaH($Na_{24}H_{24}$) by heating it from 300 K to 1800 K using Berendsen thermostat[17]. Velocity Verlet algorithm with a timestep of 0.25fs was used in all simulation runs. The heating rate was set at 0.025 K/ps(2.5×10^{10}K/s). The cluster was first minimized to find the nearest metastable state and then equilibrated at 300 K for 5×10^4 steps. The temperature of the system was then increased linearly from 300 K to 1800 K as

$$T(t) = T_{300K} + \lambda t \tag{5}$$

where λ is the heating rate.

Figure 4 shows the time evolution of the potential energy landscape(PES) during a heating simulation of NaH from 300 K to 1800 K. We notice that at 10.275 ps and 15.75 ps there are drops in PES which, is attributed to the release of H_2. Thus the conformation immediately before the release of H_2 was enthalpically unstable with respect to that at the release of H_2. This instability is tied to break up of chemical bond, whereby the two hydrogen atoms break away from NaH cluster and in the process form hydrogen molecule. When a bond breaks, its chemical potential energy is converted into kinetic energy leading to temperature rise. This temperature rise leads to increased atomic motion. The desorbed hydrogen molecule uses this energy to move away from the cluster surface.

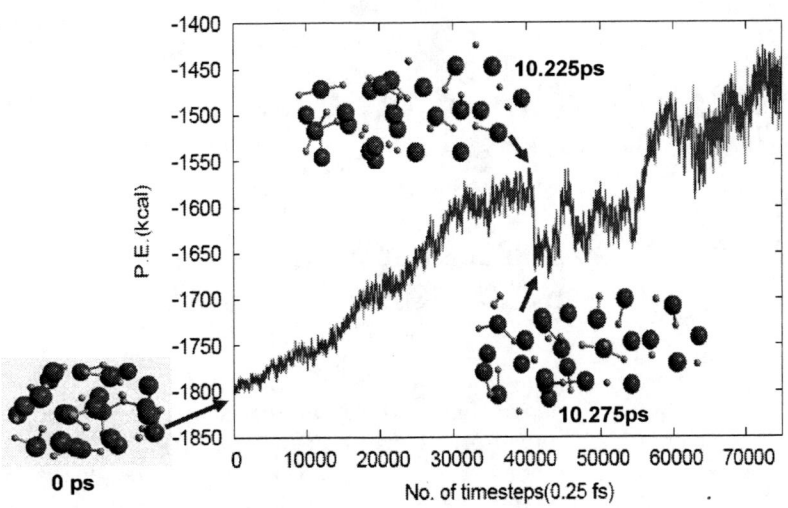

0 ps

FIGURE 4. Conformational potential energy landcape during heating of NaH cluster from 300K to 1800 K.

CONCLUSION

We have parameterized a reactive force field, ReaxFF$_{NaH}$, for Na and NaH systems by using DFT derived values for bond dissociation profiles, charge distribution, reaction energy data for small clusters and equations of state for Na and NaH condensed phases. ReaxFF$_{NaH}$ has been extensively used to sample the phase space and understand the dynamics governing hydrogen desorption during the thermal decomposition of NaH. During the heating run in molecular dynamics simulation, the signature of hydrogen desorption was a fall in the potential energy surface. This sets us on a firm footing towards parameterizing a force field for NaAlH$_4$.

ACKNOWLEDGMENTS

This work is part of the research programs of Advanced Chemical Technologies for Sustainability(ACTS), which is funded by Nederlandse Organisatie voor Wetenschappelijk Onderzoek(NWO).

REFERENCES

1. B. Bogdanovic and M. Schwickardi. *J. Alloys and Cmpds*, 253, 1997.
2. A. Strachan, A. C. van Duin, D. Chakraborty, S. Dasgupta, and W. A. Goddard. *Phys. Rev. Lett.*, 91(9):098301–+, 2003.
3. A. C. T. van Duin, A. Strachan, S. Stewman, Q. Zhang, X. Xu, and W. Goddard III. *J. Phys. Chem. A*, 107:3803, 2003.
4. Q. Zhang, T. Çağın, A. C. T. van Duin, W. A. Goddard, Y. Qi, and L. G. Hector. *Phys. Rev. B*, 69(4):045423–+, 2004.
5. A. C. T. van Duin, S. Dasgupta, F. Lorant, and W. Goddard III. *J. Phys. Chem. A*, 105:9396, 2001.
6. Sam Cheung, Wei-Qiao Deng, A. C. T. van Duin, and W. Goddard III. *J. Phys. Chem. A*, 109:851–859, 2005.
7. J. Tersoff. *Phys. Rev.*, 61:2879, 1988.
8. D. W. Brenner. *Phys. Rev. B*, 42:9458–9471, 1990.
9. M. Z. Bazant and E. Kaxiras. *Phys. Rev. Lett.*, 77:4370–4373, 1996.
10. W. J. Mortier, S. K. Ghosh, and S. J. Shankar. *J. Am. Chem. Soc.*, 120:2641, 1998.
11. G. Kresse and J. Furthmüller. *Phys. Rev. B*, 54:11169–11186, 1996.
12. P. E. Blöchl. *Phys. Rev. B*, 50:17953–17979, 1994.
13. J. P. Perdew, J. A. Chevary, S. H. Vosko, K. A. Jackson, M. R. Pederson, D. J. Singh, and C. Fiolhais. *Phys. Rev. B*, 46:6671–6687, 1992.
14. J. P. Perdew, K. Burke, and Y. Wang. *Phys. Rev. B*, 54:16533–16539, 1996.
15. J. P. Perdew, K. Burke, and M. Ernzerhof. *Phys. Rev. Lett.*, 77:3865–3868, 1996.
16. R. Dovesi, M. Causa', R. Orlando, C. Roetti, and V. R. Saunders. *J. Chem. Phys.*, 92:7402–7411, 1990.
17. H. J. C. Berendsen, J. P. M. Postma, W. F. van Gunsteren, A. Dinola, and J. R. Haak. *J. Chem. Phys.*, 81:3684–3690, 1984.

Theoretical Calculations of the Characteristics of Precious Metal Clusters

Mitsutaka Okumura[1,4], Yasutaka Kitagawa[1], Takashi Kawakami[1], Masatake Haruta[2,4] and Kizashi Yamaguchi[3]

[1] Graduate School of Science, Osaka University, 1-1 Machikaneyama, Toyonaka, Osaka, 560-0043, Japan
[2] Department of Applied Chemistry, Graduate School of Urban Environmental Sciences, Tokyo Metropolitan University,1-1 Minami-osawa, Hachioji 192-0397, Tokyo, Japan
[3] Center for Quantum Science and Technology under Extreme Conditions, Osaka University, 1-3 Machikaneyama, Toyonaka, Osaka, 560-8531, Japan
[4] Core Research for Environmental Science and Technology (CREST), Japan Science and Technology Agency, Kawaguchi-shi, Saitama 332-0012, Japan

Abstract. It is reported that core/shell type Pd/Pt bimetallic nanoclusters where the inner atoms of a Pd cluster are substituted by Pt atoms have extremely enhanced catalytic activity for cyclooctadiene hydrogenation. In order to discuss the electronic states of core/shell clusters, DFT calculations were carried out for Au_{13}, Pd_{13}, Pt_{13}, Pt/Pd_{12}, Pd/Pt_{12}, Pd/Au_{12}, Pd_{38}, and Pd_6/Pt_{32} clusters. From these calculations, it was found that the charge transfer between the core atoms and the shell atoms played an important role in the modification of the electronic state of the surface atoms.

Keywords: DFT; core/shell cluster; heterojunction; charge transfer; nanoparticle.
PACS: 73.22.-f, 71.15.Nc

INTRODUCTION

Lately, precious metal nanoparticles have received much attention as electronic devices, spintronics materials, catalysts, etc[1]. In the case of catalysis, the catalytic activities of Pd/Pt bimetallic clusters for hydrogenation were changed depending on its constituent parts and the structures of the nanoparticles. Especially, the core/shell type Pd/Pt bimetallic cluster exhibit extremely high catalytic activity for cyclooctadiene partial hydrogenation[2]. This result suggests that the characteristics of the core/shell type bimetallic clusters are dramatically changed from the catalytic point of view.

It is also well known that Au nanoparticles showed high catalytic activities for many catalytic reactions[3,4]. It is an interesting issue to modify the electronic state of Au clusters in order to increase its catalytic activity for several catalytic reactions. Therefore, Au containing core/shell type bimetallic clusters are also examined.

In order to discuss the characteristics of the bimetallic core/shell clusters, ab initio DFT calculations for N_{13}, $N-M_{12}$, (N,M=Pd, Au or Pt), and and Pd_{38}, Pt_6-Pd_{32} model systems were examind as a first step for understanding the catalytic reactions over core/shell cluster catalysts.

COMPUTATIONAL METHOD

Unrestricted hybrid DFT (UB3LYP) calculations were first carried out for the model clusters. LANL2DZ basis set was used for Au, Pd and Pt atoms. All the geometries of cubooctahedral (Coh) M_{13} models systems (M=Au, Pt and/or Pd) were optimized with O_h symmetry. The geometries of cubooctahedral (Coh) M_{38} model systems (M=Pd and Pt) were fully optimized with O_h symmetry using Dmol[3]. For these optimizations, PW91 method and DNP basis set were used. The Mulliken charges of each atom were calculated by the Mulliken population analysis. All these calculations except for the geometry optimization of M_{38} clusters were carried out by the Gaussian program[5].

RESULTS AND DISCUSSION

Characteristics of the investigated M_{13} (Pd and/or Pt) clusters

The four different clusters, such as Pd_{13}, Pt_{13}, $Pd-Pt_{12}$ and $Pt-Pd_{12}$, were investigated first. The model structures of them were shown in Figure 1. The ground spin states, total energies and the expectation values of S^2 for these clusters were listed in Table 1. From the calculation results, it was found that the ground spin states of $Pt-Pd_{12}$ and $Pd-Pt_{12}$ core/shell clusters were septet and singlet while those of Pd_{13} and Pt_{13} were triplet and septet. The calculation results also suggested that the ground spin state was changed depending on the cluster configuration. Additionally, it was also found that the different spin states and the ground spin state of each M_{13} cluster were nearly degenerate while the results were not described in detail. In Table 2, the occupation numbers of the natural orbital analysis for M_{13} and $N-M_{12}$ model clusters were summarized. The results of natural orbital analysis for these clusters show that the fractional occupation numbers were presented for all model clusters. Moreover, it was found that the sum of the number of singly occupied orbitals and natural orbitals that had a fractional occupation number for these clusters were almost same. Therefore, the large spin polarization was presented for all M_{13} model clusters. The Mulliken charges of Pt_{13} listed in Table 3 show that the surface atoms are negatively charged while those of Pd_{13} are slightly charged negatively. Therefore, it could be presumed that the charge polarization was a typical characteristic of metal clusters. Moreover, the sign of averaged surface charge density was changed when the core atom of Pd_{13} cluster was replaced with platinum. On the other hand, surface negative charge densities increased when the core atom of Pt_{13} cluster was replaced with palladium. These are also due to the charge polarization effect of M_{13} clusters induced by the charge transfer from a palladium atom to a platinum atom. In Figure 2, the difference-charge density between core/shell M_{13} model cluster systems and the monometallic M_{13} model cluster systems, superimposed on the atomic configuration are also depicted. Charge depletion and charge accumulation are shown in black and gray, respectively. These figures suggested that the charge densities were localized between the core and surface heterojunction regions of the M_{13} clusters. From these results, it was proved that the charge polarization induced by the heterojunction in core/shell structure modified the surface electronic state of the M_{13} clusters.

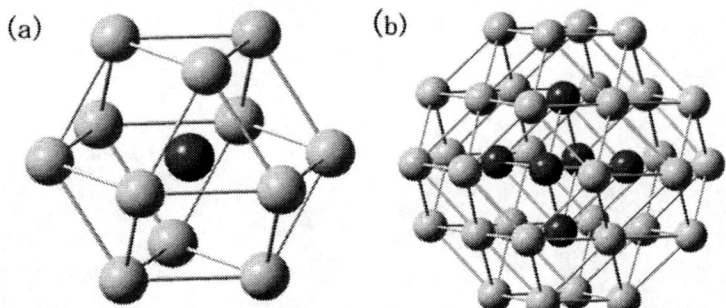

FIGURE 1. Calculated structures of model clusters: (a) M_{13} and $N-M_{12}$, (b) M_{38}, N_6-M_{32} (N, M = Au, Pd, and Pt).

TABLE 1. The calculated characteristics of M_{13} model clusters

System	Spin State	R(core-surface), Å	E_c^a, kcal/mol	$<S^2>$
Pd_{13}	triplet	2.767	462.9	3.970
Pt/Pd_{12}	septet	2.776	497.2	12.920
Pt_{13}	septet	2.754	754.1	12.162
Pd/Pt_{12}	singlet	2.749	732.9	3.561

a E_c (condensation enery)=E(Cluster)-E(sum of atomic energy in the cluster)

Characteristics of the core shell $Pd-Au_{12}$ cluster

As shown in Table 3, the surface Mulliken charge densities on the exposed surface Au atoms are much larger than those of Pd and Pt monometallic and bimetallic clusters. In the previous work, we suggested that the negatively charged surface Au atoms in Au clusters played an important role to activate the adsorbed oxygen molecules on its surface[6,7]. Therefore, the modification of the surface charge densities on Au clusters is an interesting subject to increase the catalytic activity of Au catalysts. In the next, the core shell type $Pd-Au_{12}$ cluster shown in Figure 1 was also investigated in order to examine the modification of the surface negative charge densities on Au clusters. The calculation results showed that the surface Mulliken charge densities on the exposed surface Au atoms in core/shell type $Pd-Au_{12}$ cluster are increased. This is also due to the charge polarization effect of M_{13} clusters induced by the charge transfer from a palladium atom to gold atoms. From these results, it was found that the charge polarization induced by the heterojunction in core/shell structure also modified the surface electronic state of the Au_{13} clusters.

TABLE 2. Occupation numbers of the examined M_{13} model clusters

Orbital number	Pd_{13} (triplet)	$Pt-Pd_{12}$ (septet)	Pt_{13} (septet)	$Pd-Pt_{12}$ (singlet)
122	0.01132	0.00549	0.02099	0.08012
121	0.31656	0.63321	0.02714	0.11808
120	0.38136	1.00000	1.00000	0.86250
119	0.50686	1.00000	1.00000	0.87258
118	1.00000	1.00000	1.00000	0.88774
117	1.00000	1.00000	1.00000	1.11226
116	1.49314	1.00000	1.00000	1.12742
115	1.61864	1.00000	1.00000	1.13750
114	1.68344	1.36679	1.97286	1.88192
113	1.98868	1.99451	1.97901	1.91988

TABLE 3. Mulliken charge densities of M_{13} and M_{38} model culsters

atom	Pd_{13}	$Pt-Pd_{12}$	Pt_{13}	$Pd-Pt_{12}$	Au_{13}	$Pd-Au_{12}$
core	0.827	-0.059	2.731	3.353	5.481	6.185
average of shell	-0.069	0.052	-0.228	-0.279	-0.457	-0.515

atom	Pd_{38}		Pt_6-Pd_{32}	
core	1.970		0.232	
avg. of surface edge	-0.345		0.383	
avg. of surface center	-0.378		-0.187	

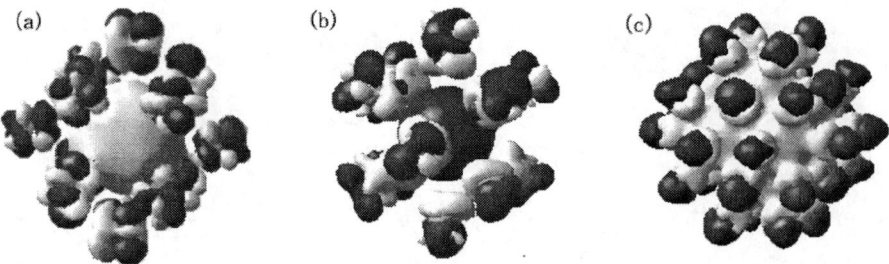

FIGURE 2. A schematic diagram of Differential charge density between (a)$Pt-Pd_{12}$ and Pd_{13}, (b) $Pd-Pt_{12}$ and Pt_{13}, and (c) Pt_6-Pd_{32} and Pd_{38}

Characteristics of the investigated M_{38} (Pd and/or Pt) clusters

In the next, monometallic Pd_{38} and core/shell type bimetallic Pt_6Pd_{32} clusters were examined. The calculated Mulliken charge densities are listed in Table 3. The Mulliken charge density of Pd_{38} indicates that the surface Pd atoms of Pd_{38} present negative charges and the absolute values of average surface negative charges were larger than those of Pd_{13} clusters. On the other hand, it was also found that the average Mulliken charges of the center atoms in the surface of core/shell Pt_6Pd_{22} become positive while the surface atoms in the edge sites of it had smaller negative charges than those of Pd_{38}. These results showed that the replacement of the core Pd atoms with Pt atoms in the M_{38} model clusters affect its surface Mulliken charge distributions. Moreover, these results also suggest that the characteristics of the molecular adsorption onto the center atoms of the exposed surfaces in core/shell Pt_6-Pd_{32} cluster would be completely different from that in monometallic Pd_{38}. It could be presumed that these results show the totally different characteristics from alloy clusters as the charge polarization mainly occurs between the neighboring hetero atoms.

In Figure 2, the difference-charge density between core/shell Pt_6-Pd_{32} model cluster system and the monometallic Pd_{38} model cluster system, superimposed on the atomic configuration are also depicted. Charge depletion and charge accumulation are shown in black and gray, respectively. From these figures, it was concluded that the charge densities were also localized between the core and surface heterojunction regions of the M_{38} cluster. These results indicated the charge polarization of M_{38} model clusters are much larger than that of M_{13} model clusters. Therefore, it could be concluded that the charge polarization effect of the clusters is size sensitive and is directly related to the heterojunction between two different metal phases in core/shell clusters. Furthermore, it was deduced that the characteristics vary strongly by changing the thickness of the outer shell and the size of the core.

CONCLUSIONS

M_{13} and M_{38} model cluster systems are examined using hybrid DFT calculations. These results suggest that the surface charge density distribution of the core/shell model clusters greatly changed from the corresponding monometallic model cluster systems. These results are due to the charge polarization induced by the charge transfer from a palladium atom to a platinum atom in the heterojunction sites of core/shell model clusters. Therefore, it was found that hetero-junction in core/shell cluster plays an important role to modify the electronic state of the cluster surface.

ACKNOWLEDGEMENTS

This work was supported by Grant-in-Aid for Scientific Research on Priority Areas (No. 19028040, "Chemistry of Concerto Catalysis") from Ministry of Education, Culture, Sports, Science and Technology, and JST, Japan

REFERENCES

1. G. Schmid, *Clusters and Colloids from Theory to Applications*, edited by G. Schmid, Weinheim, VCH, 1994.
2. N. Toshima, et al., *J. Phys. Chem.* **95**,7448-7453 (1991).
3. M. Okumura, S. Tsubota, and M. Haruta, *Cat. Lett.* **51**, pp. 53-58(1998).
4. M. Okumura, S. Tsubota, and M. Haruta, *J. Mol. Catal. A-Chem.* **199**, pp. 73-84(2003).
5. Gaussian 03, Revision C.02, M. J. Frisch, G. W. Trucks, H. B. Schlegel, G. E. Scuseria, M. A. Robb, J. R. Cheeseman, J. A. Montgomery, Jr., T. Vreven, K. N. Kudin, J. C. Burant, J. M. Millam, S. S. Iyengar, J. Tomasi, V. Barone, B. Mennucci, M. Cossi, G. Scalmani, N. Rega, G. A. Petersson, H. Nakatsuji, M. Hada, M. Ehara, K. Toyota, R. Fukuda, J. Hasegawa, M. Ishida, T. Nakajima, Y. Honda, O. Kitao, H. Nakai, M. Klene, X. Li, J. E. Knox, H. P. Hratchian, J. B. Cross, V. Bakken, C. Adamo, J. Jaramillo, R. Gomperts, R. E. Stratmann, O. Yazyev, A. J. Austin, R. Cammi, C. Pomelli, J. W. Ochterski, P. Y. Ayala, K. Morokuma, G. A. Voth, P. Salvador, J. J. Dannenberg, V. G. Zakrzewski, S. Dapprich, A. D. Daniels, M. C. Strain, O. Farkas, D. K. Malick, A. D. Rabuck, K. Raghavachari, J. B. Foresman, J. V. Ortiz, Q. Cui, A. G. Baboul, S. Clifford, J. Cioslowski, B. B. Stefanov, G. Liu, A. Liashenko, P. Piskorz, I. Komaromi, R. L. Martin, D. J. Fox, T. Keith, M. A. Al-Laham, C. Y. Peng, A. Nanayakkara, M. Challacombe, P. M. W. Gill, B. Johnson, W. Chen, M. W. Wong, C. Gonzalez, and J. A. Pople, Gaussian, Inc., Wallingford CT, 2004.
6. M. Okumura, Y. Kitagawa, M. Haruta and K. Yamaguchi, *Chem. Phys. Lett.* **346**, pp163-168 (2001).
7. M. Okumura, Y. Irie, Y.Kitagawa, T. Fujitani, Y. Maeda, T. Kasai,and K. Yamaguchi, *Catalysis Today,* **111**, pp. 311-315 (2006)

Charge Transport in Conjugated Materials: From Theoretical Models to Experimental Systems

Yoann Olivier[1], Luca Muccioli[2], Claudio Zannoni[2], and Jérôme Cornil[1]

[1]*Laboratoire de Chimie des Matériaux Nouveaux, Université de Mons-Hainaut,*
20, Place du Parc, 7000 Mons, Belgium
[2]*Dipartimento di Chimica Fisica e Inorganica, Università di Bologna,*
4, Viale Risorgimento, 40136 Bologna, Italy

Abstract. Charge carrier mobility is the key quantity to characterize the charge transport properties in devices. Based on earlier work of Bässler and co-workers, we set up a Monte-Carlo approach that allows us to calculate mobility using transfer rates derived from Marcus theory. The parameters entering into the rate expression are evaluated by means of different quantum-chemical techniques. Our approach is applied here to a model one-dimensional system made of pentacene molecules as well as to real systems such as crystalline structures and columnar liquid crystal phases.

Keywords: Charge transport, Marcus theory, quantum-chemistry, Monte-Carlo simulations.
PACS: 73.23.–b, 72.20.Ee

INTRODUCTION

Since the discovery of high electrical conductivity in organic conjugated polymers, the interest dedicated to organic semiconductors as active elements in opto-electronic devices such as field-effect transistors (FETs), light-emitting diodes (LEDs) and solar cells, is constantly growing. Organic FETs are made of three metallic contacts (gate, drain, and source); their operation consists in modulating the current flowing from the source to the drain across the organic layer by means of a voltage applied at the gate [1]. Organic solar cells [2] and LEDs [3] are built by sandwiching an organic layer between two electrodes and aim at converting an incident light into charges and at creating light upon charge injection, respectively. The performance of these devices strongly rely on the charge transport properties of the materials (small conjugated molecules or polymers) [4].

From a macroscopic point of view, the charge transport properties are quantified by the hole and/or electron mobility. The mobility μ represents the ease of charge migration across a layer of conjugated materials and is defined as:

$$\mu = \frac{d}{t \cdot \left| \vec{F} \right|} \tag{1}$$

where d is the distance travelled by the charge carrier, t the time to travel that distance and $\left| \vec{F} \right|$ the norm of the electric field applied across the device.

METHODOLOGY

In order to optimize the performance of the devices, it is important to uncover the way to maximize the mobility of the charge carriers in the organic layer to allow for a high ON/OFF ratio and a high switching frequency in FETs, an efficient charge recombination in LEDs and charge separation in solar cells. This challenging task can be

CP1046, *Selected Papers from ICNAAM 2007 and ICCMSE 2007*
edited by T. E. Simos, G. Maroulis, G. Psihoyios, and Ch. Tsitouras

accomplished by combining experimental measurements on "simple" systems (e.g. perfect crystals), quantum chemistry calculations of the relevant molecular properties and simulation techniques to reproduce both morphologies and charge dynamics.

To this end, in the early nineties Bässler and co-workers [5] studied intensively disordered organic systems (doped or non-doped) and developed Monte-Carlo methods to simulate charge transport within a hopping regime in a model cubic lattice. The approach was based on effective parameters to define the cubic lattice constant (which is the average distance between adjacent sites) and the Miller-Abrahams transfer rates, introducing both energetic and positional disorder without taking into account exact chemical structures.

In the recent years, our group developed a quantum-chemical approach to get a deeper insight into charge hopping processes at the molecular scale [6]. In this framework, the hopping rate is expressed within the semi-classical Marcus theory that implies passing through a transition state when going from the reactants to the products. The transfer integral (t) and internal reorganization energy (λ_i) that appear in the Marcus formula are evaluated with quantum-chemical calculations.

In a next step, we have developed a Monte-Carlo algorithm to estimate charge mobilities [6] on the basis of the transfer rates obtained via the Marcus theory and related approaches for each pair of interacting molecules [7].

SYSTEMS OF INTEREST

We first applied this methodology to model one-dimensional systems made of face-to-face pentacene molecules, to investigate the influence of the amplitude of the electric field, the amplitude of the reorganization energy, positional disorder (translation parallel and perpendicular to the stacking axis, gaussian distribution of intermolecular distances) and energetic disorder (presence of traps).

Our simulation points to a Poole-Frenkel behaviour of the mobility in a large range of electric fields and demonstrate that there is no direct relationship between the degree of spatial overlap between adjacent molecules and the charge mobility values (see **FIGURE 1**).

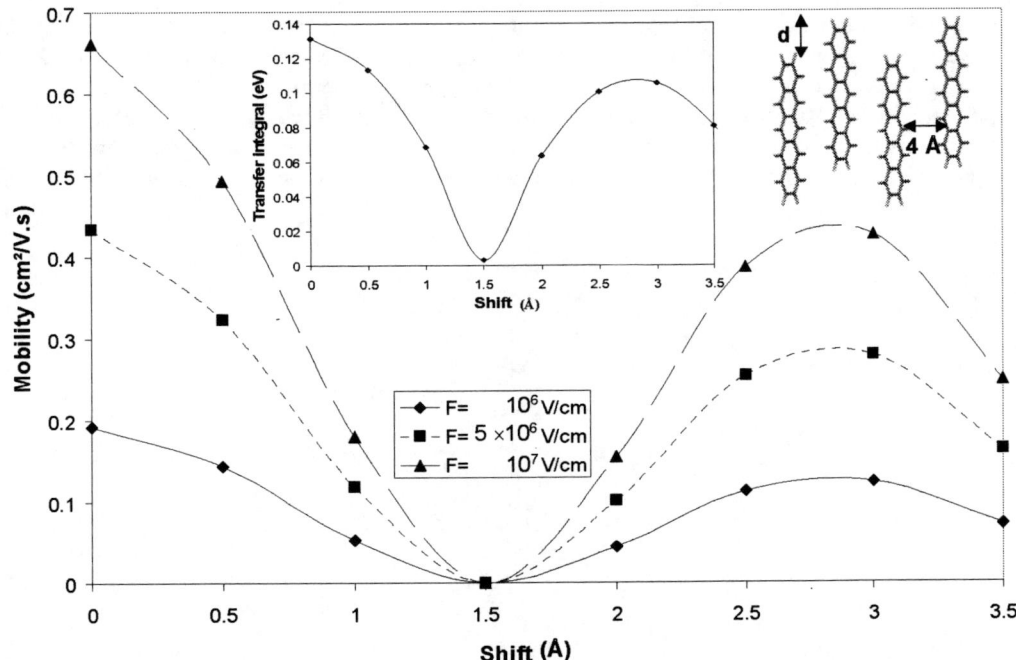

FIGURE 1. Evolution of the hole mobility as a function of the shift applied to every single molecule over two in a direction perpendicular to the stacking axis in a one-dimensional array of pentacene molecules separated by 4 Å. The inset shows the evolution of the corresponding transfer integral.

We have extended this approach to systems with a higher dimensionality, such as crystalline structures, in particular that of rubrene in view of the numerous experimental studies performed recently. A main objective was to rationalize the mobility anisotropy reported in the layer of the rubrene single crystal by Podzorov and co-workers [9]. Our simulation of the mobility anisotropy based on a purely hopping regime yields similar trends compared to experiment, with a mobility minimum and maximum along the a axis and b axis of the rubrene crystalline structure, respectively (see **FIGURE 2**). Moreover, the range of mobility values obtained is in good agreement with the OFET mobilities extracted from the linear regime.

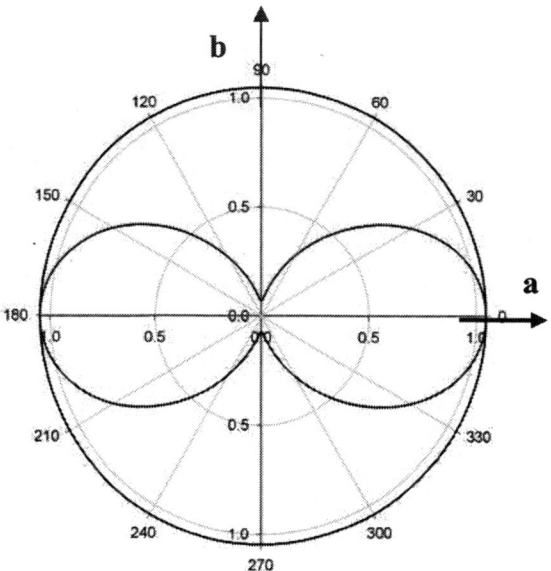

FIGURE 2. Hole mobility anisotropy in the rubrene single crystal obtained from MC simulations based DFT transfer integrals (F=1000 V/cm; λ_s = 0.15 eV).

Recently, Troisi and Orlandi [9] showed that lattice vibrations strongly modulate the transfer integrals in their calculations focusing on pentacene and anthracene molecular clusters. They combined in their approach force-field molecular dynamics simulations to record molecular trajectories and the semi-empirical Hartree-Fock technique INDO (Intermediate Neglect of Differential Overlap) to compute the transfer integrals along the trajectory. We have applied the same combination of computational techniques to columns of metal-free phthalocyanines (see Figure 3) to study the influence of the dynamics of positional disorder and orientational disorder on the mobility values.

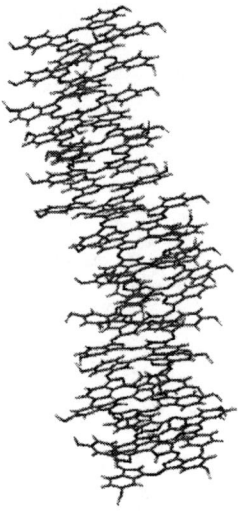

FIGURE 3. Snapshot of a column of phthalocyanine molecules

ACKNOWLEDGMENTS

The work in Mons is partly supported by the Belgian Federal Governement "Interuniversity Attraction Pole in Supramolecular Chemistry and Catalysis, PAI 5/3"; Région Wallonne (Project ETIQUEL); the European Integrated project NAIMO (NMP4-CT-2004-500355); and the Belgian National Fund for Scientific Research (FNRS/FRFC). J.C. is an FNRS Research Fellow; Y.O. acknowledges a grant from "Fonds pour la Formation a` la Recherche dans l'Industrie et dans l'Agriculture (FRIA)". The work in Bologna is funded by the MIUR national project "Modelling and characterisation of liquid crystals for nano-organised structures (PRIN #2005035119) and by the European Integrated project NAIMO.

REFERENCES

1. G. Horowitz, *Adv. Mater.* **10**, 365-377 (1998).
2. C.J. Brabec, N.S. Sariciftci, J.C. Hummelen, *Adv. Funct. Mater.* **11**, 15-26 (2001)
3. J. H. Burroughes, D. D. C. Bradley, A. R. Brown, R. N. Marks, K. Mackay, R. H. Friend, P. L. Burns, A. B. Holmes, *Nature* **347**, 539-541 (1990)
4. V. Coropceanu, J. Cornil, D. A. da Silva Filho, Y. Olivier, R. Silbey, J. L. Brédas, *Chem. Rev.* **107**, 926-952 (2007)
5. Bassler, H. *Phys. Status Solidi B*, **175**, 15-56 (1993)
6. J. L.Bredas, D. Beljonne, V. Coropceanu, J. Cornil, *Chem. Rev.* **104**, 4971-5004 (2004)
7. Y. Olivier, V. Lemaur, J.L. Brédas, J. Cornil., *J. Phys Chem* A **110**, 6356-6364 (2006)
8. J. Jortner, *J. Chem. Phys.* **64**, 4860-4867 (1976)
9. V. C. Sundar, J. Zaumseil, V. Podzorov, E. Menard, R. L. Willett, T. Someya, M. E. Gershenson, J. A. Rogers, *Science* **303**, 1644-1646 (2004).
10 A. Troisi, G. Orlandi, *J. Phys Chem A* **110**, 4065-4070 (2006)

Modeling of Frenkel vs. CT Excitons – Solved and Unsolved Problems

Piotr Petelenz

K. Gumiński Department of Theoretical Chemistry, Jagiellonian University
Ingardena 3, 30-060 Cracow, Poland

Abstract. The difficulties that emerge in modeling the interaction between the Frenkel and CT excitons in organic molecular systems are reviewed, with special emphasis on vibronic coupling problems. Special attention is given to molecular crystals with very intense Frenkel exciton transitions, such as oligothiophenes.

Keywords: Frenkel excitons, CT excitons, polaritons, electro-absorption, vibronic coupling.
PACS: 71.35.Cc, 71.36.+c, 78.40.Me, 78.90.+t

INTRODUCTION

Organic molecular crystals, viewed as potential components of optoelectronic devices, elicit considerable interest in recent literature. In this context, it is of paramount importance to understand the properties of their excited electronic states, especially those of charge-transfer (CT) origin (relevant in their capacity of charge-carrier precursors). In general, any electronic excitation of the crystal may be represented as a combination of Frenkel states, where the electron and the hole reside at the same molecule, and CT configurations, where the two charges reside at different molecules. The relative contents of the two kinds of configurations determine the properties of the excited state in hand; in principle this information may be readily obtained by diagonalizing the appropriate Hamiltonian matrix. This is easier said than done, since the approximations needed to describe adequately one of these two manifolds are sometimes hard to reconcile with those required to describe the other one. This situation is aggravated when vibronic coupling has to be included, which is usually the case. The objective of the present paper is to discuss the emerging difficulties.

CHARGE-TRANSFER EXCITONS

Many properties of charge-transfer configurations, where the electron and the hole are located at different molecules, are governed by intermolecular overlap. Their intrinsic transition dipole moments from the ground state are overlap-limited, and so are the matrix elements that couple the CT states to the Frenkel manifold. Owing to the relatively short range of molecular orbitals, the spectroscopically relevant CT configurations are only those that involve closely spaced molecules; the model may be usually limited to the nearest and second-nearest neighbors. When the needed parameters (CT integrals, Frenkel exciton dissociation integrals) are calculated by quantum chemistry methods, the main problem consists in using a basis set that correctly describes the long-range tails of the wavefunctions. In order to parametrize the diagonal energies of CT configurations, one needs the ionization potential and electron affinity of the constituent molecules, which may be obtained either from calculations or from experiment, and the electrostatic stabilization energies, readily obtainable from microelectrostatic calculations. The eigenstates of CT origin are also sensitive to the energies of Frenkel excitons owing to their coupling to these latter states via the dissociation integrals; this gives rise to nontrivial directional dependences (*vide infra*).

The coupling of CT states to intramolecular vibrations results from the change of the equilibrium position of the nuclei upon ionization. This effect was usually treated in the limit of strong vibronic coupling, where the vibrational excitation is assumed to always accompany the electronic excitation, which is justified, since the splittings in the CT manifold are relatively small, being limited by intermolecular overlap. In this approach, the contributions that

CP1046, *Selected Papers from ICNAAM 2007 and ICCMSE 2007*
edited by T. E. Simos, G. Maroulis, G. Psihoyios, and Ch. Tsitouras
© 2008 American Institute of Physics 978-0-7354-0574-5/08/$23.00

represent separated electronic and vibrational excitations are entirely disregarded. These vibronic states are expected to split off the tight-binding exciton-phonon configurations; the splitting is probably too small to be experimentally resolved, hence no obvious deviations from this approximate description were detected so far.

FRENKEL EXCITONS VS POLARITONS

The properties of Frenkel excitons, where the electron and the hole reside at the same molecule, are insensitive to intermolecular overlap. This applies especially to the transition dipole moments from the ground state, which may be very large. If this is the case, the excitations of this kind exhibit some peculiarities that need careful theoretical handling.

1. The lattice sums of the intermolecular resonance integrals that define the dispersion relations and Davydov splittings are only conditionally convergent and in effect are singular at the centre of the Brillouin zone, so that the exciton energy at that point strongly depends on the direction of the wavevector. This effect (giving rise to the splitting between transverse and longitudinal excitons) is due to the long-range part of the interaction and cannot be easily included even in sophisticated cluster models (such as [1]). Owing to the coupling via the dissociation integrals, this directional dependence spills onto the eigenstates of CT provenance.

2. Very intense excitons strongly interact with photons. The dispersion relation for the ensuing entity called a polariton exhibits a gap, experimentally observed as a band of metallic reflection (stopping band). Until recently, the energetic consequences of this interaction were often exaggerated; in reality they are rather marginal [2].

3. Many anomalies commonly attributed to polariton effects are in fact due to the peculiarities of vibronic coupling; the latter results from the change of the nuclear equilibrium positions upon electronic excitation. Its description that may be considered classic in this context, has two versions [3], both deriving from the limit of strong vibronic coupling, similar to that mentioned above for the CT states. In this approximation, the vibrational excitation is assumed to reside at the same molecule as the electronic excitation and to perfectly accompany the latter in its propagation; conveniently, each vibrational state (characterized by the quantum number v) individually splits into Davydov components. This approach is applicable as long as the resonance interaction between the molecules is not too large with respect to vibronic stabilization (i.e. to the nuclear relaxation energy upon electronic excitation of the molecule). It works reasonably well even for moderately intense electronic transitions [3]. However, for very intense Frenkel states the electronic interaction between the molecules is strong (as manifested in the large Davydov splitting) and clearly prevails over the vibronic term. In that case, the approach rooted in the strong-vibronic-coupling limit is no longer valid, since the actual wavefunction is bound to contain substantial contributions from the configurations where the vibrational excitation is separated from the electronic excitation. When forcibly applied beyond its range of applicability, the classic approximation predicts a seemingly anomalous intensity redistribution between the vibronic states, sometimes erroneously attributed to polariton effects [2-4]. An enormous peak, located either at the low- or high-energy end (depending on the lattice sums), dominates the absorption spectrum. This peak corresponds to the 0-0 line of the purely electronic transition and could be easily obtained in the opposite (Born-Oppenheimer) limit of *weak* vibronic coupling, appropriate in this case [3]; in reality, the apparent anomaly results from the breakdown of the strong-vibronic-coupling approximation. In addition, the dominant band has a peculiar asymmetric shape, with long low-energy onset and a steep high-energy cut-off, due to the admixture of unbound states from the exciton-phonon continuum [5].

MIXING BETWEEN FRENKEL AND CT CONFIGURATIONS – SPECTROSCOPIC CONSEQUENCES

As it has already been mentioned, the actual crystal eigenstates are combinations of Frenkel and CT configurations. In most cases the contribution from one of the manifolds prevails, which facilitates the interpretation.

Owing to their large transition dipole moments, the eigenstates of Frenkel parentage dominate in absorption spectroscopy. The overlap-limited intrinsic CT transition dipoles are much smaller, so that the eigenstates of CT provenance borrow a substantial part of their intensity from intramolecular excitations, the borrowing being mediated by Frenkel exciton dissociation integrals. The resultant absorption is weak anyway and is usually masked by the vibronic replicas of Frenkel transitions. There are some exceptions: owing to a peculiar arrangement of the molecules in the unit cell, the *b*-polarized Davydov component in oligothiophene crystals has a negligible transition moment [6], which exposes the (generally small) contribution from the eigenstates of CT origin [7,8].

The method of choice for experimental studies of CT states is electro-absorption (EA) spectroscopy, where their large sensitivity to electric field compensates their small absorption intensity, producing signals comparable to those of Frenkel excitons. Although in the eigenstates of centrosymmetric crystals the large permanent dipole moments of individual CT configurations add up to zero [9], in the symmetry-adapted basis they give rise to (equally large) off-diagonal dipole moments, and these in electric field mediate an interaction between the zero-field eigenstates. The shape of the resultant EA signal depends on the zero-field splitting between the thereby coupled eigenstates. When the splitting is large with respect to the optical width, each of the two coupled states that are pushed apart by the dipolar perturbation gives rise to a first-derivative EA signal (as observed e.g. for the low-energy CT manifold of the fullerene crystal [10]). When the splitting is small with respect to the optical width (which is the most common situation), the two first-derivative signals (one simple and one inverted) fuse to yield a second-derivative EA profile [9], as predicted by classic analysis [11].

For a Frenkel state, the shift induced by the electric field is due to the change of the molecular polarizability upon electronic excitation, is quadratic in the field strength, and is much smaller. It always yields a first-derivative EA profile [12], in contradistinction to the signals from CT excitons which may follow either the first or the second derivative of the corresponding absorption band.

Phenomenological analysis of an EA spectrum, unassisted by microscopic theoretical calculations, is difficult and likely to produce artefacts, because the signals from different states are often congested and tend to overlap, and because some signals may contain both the first- and second-derivative contributions, especially when the corresponding eigenstates have a mixed (CT+Frenkel) parentage.

VIBRONIC COUPLING

As long as vibronic effects are negligible (for instance in fullerene [10]), the technical difficulties in theoretical calculations of the EA spectrum consist in evaluating the input parameters. It should be noted in passing that just in the specific case of fullerene, the degeneracies due to the high symmetry of the molecule are an additional complication that has to be circumvented by applying a simplified model, but the issue of Frenkel-CT mixing is straightforward enough.

The situation is only somewhat more difficult in the opposite limit, i.e. when the molecular distortion parameters are large. The EA spectrum is then comfortably reproduced within the strong-vibronic-coupling approximation (e.g. in polyacenes [13]), which is only slightly more demanding numerically and its main conclusions can be stretched even to some cases which formally qualify in the intermediate-coupling regime.

The crucial difficulty arises for very intense Frenkel transitions when vibronic coupling is substantial (manifested in the absorption spectrum), but the resonance interactions are dominant nevertheless. This is the case for oligothiophenes. In order to describe exciton-phonon coupling in the CT manifold, the strong-vibronic-coupling approximation would still be desirable; however (as mentioned in the preceding section), it would inevitably produce artefacts in the Frenkel manifold, where the enormous resonance splittings render it absolutely invalid. In addition, the peculiar absorption profiles of intense high-energy Frenkel states (*vide supra*) can be rationalized only in the limit of weak vibronic coupling where the unbound exciton-phonon continuum can be explicitly represented. For the sake of consistency, the limit of weak vibronic coupling was so far used also for the CT manifold, and it turned out to be numerically quite successful (at least for polarization *b* of the sexithiophene spectrum [7]). The reasons of this success are unclear; also the spectra for the other polarization are yet to be calculated. This will necessitate a more comprehensive approach to vibronic coupling.

The standard methodology of expanding the sought vibronic/excitonic wavefunction in a large basis set of Frenkel/CT configurations for the electronic part and of the ground-state harmonic oscillator wavefunctions for the vibrational degrees of freedom is not feasible because of the prohibitive numerical effort involved, since the convergence of the corresponding expansion is much too slow. This approach is practicable for relatively small aggregates [14] but the obstacles rapidly grow with the size of the system; solutions for a large crystal are not likely to be ever attainable.

Based on experience collected in aggregate calculations [1,14], the alternative expansion in the Lang-Firsov (LF) basis where the oscillators are shifted to their excited-state equilibrium positions looks more promising in this regard. The first term of the expansion reconstitutes the wavefunction of the strong-vibronic-coupling limit, which is an interpretational advantage especially for the eigenstates of CT origin. The potential shortcoming is that the continuum states can be reproduced only in the limit of infinite basis. The other questionable point is how effective this basis will prove after the Fourier transformation, necessary for a large crystal. Studies in this direction are presently underway.

FILM SPECTRA

The difficulties are further compounded by the fact that practically all experimental EA spectra (with one exception to-date [6]) were measured not for single crystals, but for disordered films. Realistic simulation of the disorder is a major problem; fortunately, the recent progress in experimental texture studies provides useful hints.

Even if the texture of the sample on which the pertinent EA spectrum was measured were strictly known, the disorder enormously increases the computational effort because of the inevitable averaging. It has to be done numerically, since the relevant eigenstates exhibit nontrivial directional dependences due to the nonanalytic behaviour of Frenkel state energies. Besides, there is the obvious dependence on crystallite orientation with respect to the electric field, and the standard influence of crystallite orientation on absorption intensities. By and large, the space of orientational parameters has to be sampled densely enough, and this is not feasible if an individual run for a specified orientation is overly demanding numerically. This consideration highlights the preference for simple models in the EA context.

A potentially feasible strategy to build a model of this kind that would be applicable even for systems with very intense electronic transitions is inspired by the fact that the states that are the most prominent in electro-absorption spectra are usually not the same as those relevant for absorption spectroscopy. The broad bands, characteristic of strong Frenkel transitions, are dominant in absorption but barely noticeable in EA signals, owing to the derivative shape of the latter. In contrast, the eigenstates of CT provenance only rarely substantially contribute to absorption. On this view, one may hope to reasonably reproduce the EA spectrum within the approximation of strong vibronic coupling, provided that the input diagonal energies of Frenkel states are adjusted to fit within that model the actual positions of the observed absorption bands. On the other hand, the absorption spectrum is reasonably well handled in the limit of weak vibronic coupling. This approach has some obvious disadvantages. Firstly, it is not internally consistent, because for the same system essentially contradictory approximations are then invoked in different interpretational contexts, which eliminates the possibility of additional cross-checks. Secondly, the dividing line is not clear-cut, since for some systems (such as oligothiophenes) the Frenkel exciton absorption spectrum exhibits some features characteristic of the intermediate-coupling regime, while, on the other hand, the Born-Oppenheimer limit may sometimes work also for CT states [7]. Thirdly, the accuracy is inherently limited by the necessity to use the hypothetical strong-vibronic-coupling Frenkel exciton energetics to generate the CT state vibronic energies. An alternative approach, internally more consistent but numerically more difficult, is rooted in the Lang-Firsov basis and is presently being tested. For CT contributions, the first term of the expansion (equivalent to the strong-vibronic-coupling approximation) is already a good starting point. At the same time, the basis is flexible enough to work well in the intermediate-coupling region, providing reasonable Frenkel exciton energies. The continuum of unbound Frenkel-exciton-phonon states can be approximately treated in terms of the Fano model. Altogether, the approach is computationally demanding, but seems tractable, even with the additional effort of orientational averaging.

REFERENCES

1. F. C. Spano, *J. Chem .Phys.* **118**, 981-994 (2003).
2. P. Petelenz and A. Stradomska, *phys. stat. sol.(c)* **3**, 3472-3475 (2006).
3. M. Andrzejak and P.Petelenz, *Chem. Phys.* **335**, 155-163 (2007).
4. S. Möller, G. Weiser and C. Taliani, *Chem. Phys.* **295**, 11-20 (2003).
5. W. Kulig and P. Petelenz (in preparation).
6. S. Möller, G. Weiser and F. Garnier, *Phys. Rev. B* **61**, 15749-15755 (2000).
7. M. Andrzejak, P. Petelenz, M. Slawik and R. W. Munn, *J. Chem. Phys.* **117**, 1328-1335 (2002).
8. S. Tavazzi, M. Laicini, L. Raimondo, P. Spearman, A. Borghesi, A. Papagni and S. Trabattoni, *Appl. Surf. Science* **253**, 296-299 (2006).
9. P.Petelenz, Charge transfer excitons in organics, in Organic Nanostructures: Science and Applications, edited by V. M. Agranovich and G. C. La Rocca, Proceedings of the International School of Physics "Enrico Fermi", Course CXLIX, Società Italiana di Fisica, Bologna 2002, pp. 1-21.
10. B. Pac, P.Petelenz, M.Slawik and R.W.Munn, *J.Chem.Phys.* **109**, 7932-7939 (1998).
11. L. Sebastian, G. Weiser and H. Bässler, *Chem. Phys.* **61**, 125-135 (1981).
12. P. Petelenz and A. Stradomska, *Phys. Rev. B* **71**, 235205 (2005).
13. P. Petelenz, M. Slawik, K.Yokoi and M.Z.Zgierski, *J. Chem. Phys.* **105**, 4427-4440 (1996).
14. M. Andrzejak (private communication).

Calculation of Frequency-Dependent Polarizabilities and Hyperpolarizabilities Based on the Quasienergy Derivative Method (Is the Numerical Approach Impossible?)

Kotoku Sasagane

Faculty of Business Management, Takachiho University, Suginami-ku, Tokyo 168-8508, Japan

Abstract. The essence of the quasienergy derivative (QED) method and calculations of the frequency-dependent hyperpolarizabilities based on the QED method will be presented in the first half. Our recent and up-to-date development and some possibilities concerning the QED method will be explained later. At the end of the lecture whether the extension of the QED method to the numerical approach is possible or not will be investigated.

Keywords: quasienergy derivative; hyperpolarizability; electric susceptibility; response property; nonlinear optics; FVMO
PACS: 33.15.Kr, 42.65.An, 32.10.Dk

INTRODUCTION

Owing to the recent progress of laser technology, analysis of the nonlinear optical process has become of increased interest. Quantum mechanical studies of the nonlinear properties of isolated molecules are very important to explain the nonlinearity of crystals because there is a close relationship between the nonlinear electric susceptibilities of a crystal and the hyperpolarizabilities of the component molecules. The calculation of dynamic (=frequency-dependent) hyperpolarizabilities as long as static ones seems very important since there have been several experimental measurements of nonlinear electric susceptibilities for small gas-phase systems so they can be directly compared with the theory only if the theoretical evaluation of hyperpolarizabilities is frequency-dependent.

The author and the co-workers have published the availability of the quasienergy derivative (QED) method to calculation of the frequency-dependent hyperpolarizabilities [1-11]. In the first half of this lecture we will present the essence of our metodlogy and these works. The latter half will provide our recent and up-to-date development. Some possibilities concerning the QED method will be also explained.

OUTLINE OF LECTURE

At the beginning of this lecture the author would like to mention how we could hit on the QED method. Then the outlined explanation of the QED method follows. After that we present some calculations of our research group. α $(-\omega;\omega)$, β $(-\omega;\omega,0)$, β $(-2\omega;\omega,\omega)$, γ $(-\omega;\omega,0,0)$, and γ $(-2\omega;\omega,\omega,0)$ of the FH, H_2O, CO, and NH_3 molecules are calculated at the time-dependent Hartree-Fock (TDHF) and the Second-Order Møller-Plesset perturbation theory (MP2) levels [7-9]. α $(-\omega;\omega)$, γ $(-\omega;\omega,0,0)$, and γ $(-2\omega;\omega,\omega,0)$ of the Ne atom are also calculated [7, 11]. As for open-shell systems we use the spin-restricted form throughout. α $(-\omega;\omega)$ of the Li, Na, and K atoms and BeH, MgH, CaH, CN, and NH_2 molecules are calculated at the TDHF and MP2 levels [6]. $\gamma(-\omega_\sigma;\omega_1,\omega_2,\omega_3)$ of Li, Na, K, and N atoms are calculated at the TDHF level [10]. We also report the modified (Hy)POL basis set named *the ModPol series* [12], which is suitable for the calculation of (hyper)polarizabilities.

The characteristic feature of the QED method is that the derived response properties are always in exact agreement with the corresponding static properties by means of the numerical differentiation of energies with respect to the static electric field, which we hereafter call the finite-field (FF) method, at the static limit of outer field. Table

CP1046, *Selected Papers from ICNAAM 2007 and ICCMSE 2007*
edited by T. E. Simos, G. Maroulis, G. Psihoyios, and Ch. Tsitouras
© 2008 American Institute of Physics 978-0-7354-0574-5/08/$23.00

1 denotes our awareness of analytical calculations of dynamic response properties (frequency-dependent α, β, and γ) in literature in this context, where "its equivalent method" means the agreement with the FF method at the static limit. Although Table 1 is far from completeness, it can still offer some information to the readers.

TABLE 1. Analytical calculations of dynamic response properties by means of the QED method or its equivalent method, where 'Y' denotes that the analytical calculation on this level of theory has been already reported, 'W' denotes that the first calculation on this level of theory is by our research group, and '–' denotes that we can not find any calculation on this level of theory.

Level of theory	α	β	γ
Closed-shell TDHF	Y[a,b]	Y[a,b,c]	Y[a,b]
TDUHF	Y[d]	Y[e]	–
TDROHF	Y[f,g,h]	W	W
Closed-shell TDHF-MP2	W	W	–
TDROHF-MP2	W	–	–
TDMCSCF	Y[i,j,k]	Y[l,m]	Y[n]
Limited CI	–	–	–
CC (unrelaxed orbitals)	Y[o,r]	Y[p,r]	Y[q,r]
CC (relaxed orbitals)	–	–	–
TDDFT	Y[s,v]	Y[t,v]	Y[u,v]

[a]We can find several earlier calculations of small closed-shell molecules such as H_2, He, and LiH in Refs.[16-19].

[b]Sekino and Bartlett reported extensive calculations of the series of molecules CH_4, CH_3F, CH_2F_2, CHF_3, and CF_4 together with H_2 and FH in Ref.[20], which is one of the most important work in this field.

[c]Aiga *et al.* reported $\beta(-\omega_\sigma;\omega_1,\omega_2)$ of FH, which corresponds to the sum frequency generation (SFG), in Ref.[21].

[d]Karna first implemented and reported the calculations of H, C, N, O, F, Si, P, S, Cl, O_2, NO, and OH in Ref.[22].

[e]Karna first implemented and reported the calculations of NO and OH in Ref.[22].

[f] Yeager *et al.* reported TDROHF $\alpha(-\omega,\omega)$ of O_2 in contrast with their TDMCSCF results in Refs.[23,24].

[g]Hettema and Wormer reported TDROHF $\alpha(-\omega,\omega)$ of H, N, CN, NH, and OH^+ in Ref.[25], where imaginary frequencies are also used in order to obtain van der Waals coefficients.

[h] Kobayashi *et al.* reported TDROHF $\alpha(-\omega,\omega)$ of Li, Na, K, BeH, MgH, CaH, CN, and NH3 in contrast with their TDROHF-MP2 results in Ref.[6].

[i]MCTDHF(MCRPA) $\alpha(-\omega,\omega)$ of Be in Ref.[23] and O_2 in Refs.[23,24] seem the pioneering work in this context.

[j]Jaszuński and McWeeny reported the calculations of He and H2 in Ref.[26], where imaginary frequencies are also used in order to obtain van der Waals coefficients of He-He, He- H_2, and H_2- H_2 interactions.

[k]Some references are also listed in Refs.[1,27].

[l]The dc-SHG $\gamma(-2\omega,0,\omega,\omega)$ values of Ne were obtained from MCQR SHG $\beta(-2\omega;\omega,\omega)$ calculations via the FF method in Refs.[28,29].

[m]MCQR SHG $\beta(-2\omega,\omega,\omega)$ calculations were reported in Ref.[30].

[n]MCCR calculations of LiH and CO are in Ref.[31].

[o]CCLR $\alpha(-\omega,\omega)$ of LiH are in Ref.[32].

[p]CCRQ SHG $\beta(-2\omega,\omega,\omega)$ and OR $\beta(0;\omega, -\omega)$ of FH are in Ref.[33].

[q]CCCR dynamic γ of N_2 for several third-order optical processes are in Refs.[34,35] and ones of Ne, Ar, and CH_4 are in Ref.[35].

[r]Recent works are summarized in Ref.[36].

[s]Recently $\alpha(-\omega,\omega)$ of Hg, AuH, and PtH_2 at 4-component relativistic density-functional level were reported in Ref.[37].

[t]SHG $\beta(-2\omega,\omega,\omega)$ calculations of large organic molecules were reported in Ref.[38].

[u]Dynamic γ calculations for N_2, benzene, and the C60 fullerene were reported in Ref.[39].

[v]Recent works are summarized in Ref.[40].

In the latter half of the lecture, the author would like to present some unpublished works, our recent and up-to-date development, and some possibilities concerning the QED method rather than concluding remarks. In this decade the author has been interested in the fully variational molecular orbital (FVMO) theory by Kazuhide Mori and co-workers [13, 14]. At first we will present the latest version of the modified (Hy)POL basis set, which is referred to as *the OptPol series*, where all the orbital exponents are completely optimized with the atomic ROHF calculations by using the extended version of the FVMO program named GAMERA. Then we will introduce some calculated polarizabilities and hyperpolarizabilities obtained from GAMERA though all of them are static ones by means of the FF method. The extension of the QED method to the FVMO theory would be possible in principle [14] but the actual implementation of the QED method on GAMERA will be very difficult even at the TDHF level, because the programming codes must include the time developments of orbital exponents and the center positions of all basis functions in addition to the standard parameters.

At the end of the lecture we would like to refer to the possibilities concerning the QED method. The absence of the calculation (–) on Table 1 means that the implementation of the programming codes of the QED method or its equivalent method at any high level of theory is not so easy. On the other hand, the corresponding calculations of static polarizabilities and hyperpolarizabilities are always easy enough only if the codes include the simple module of the FF method. However, we know that the history of computational chemistry has proved that some analytical difficulties in physical problems can be accurately solved in numerical computations. Is the extension of the QED method to the numerical approach impossible?

ACKNOWLEDGMENTS

The author wishes to thank Prof. Kizashi Yamaguchi and Prof. Masayoshi Nakano for their continuous hospitality and their great contribution in this field. The author also would like to thank Dr. Fumihiko Aiga, Dr. Takao Kobayashi, Dr. Motoyuki Shiga, Mr. Kazuhide Mori (Waseda Computational Science Consortium), and Prof. Kazunari Suzuki (Takachiho University) for their continuous discussions and supports.

The author is also grateful to Dr. Benoît Champagne for recommending me to attend the Symposium.

REFERENCES

1. K. Sasagane, F. Aiga, and R. Itoh, *J. Chem. Phys.* **99**, 3738-3778 (1993).
2. F. Aiga, K. Sasagane, and R. Itoh, *J. Chem. Phys.* **99**, 3779-3789 (1993).
3. K. Sasagane, "The Theoretical Derivation of the Frequency-Dependent Polarizabilities and Hperpolarizabilities of Aoms and Molecules: The Presentation of the Quasienergy Derivatives Method", Dr. Thesis, Waseda University, 1994 (in Japanese).
4. F. Aiga, K. Sasagane, and R. Itoh, *Int. J. Quantum Chem.* **51**, 87-97 (1994).
5. F. Aiga and R. Itoh, *Chem. Phys. Letters* **251**, 372-380 (1996).
6. T. Kobayashi, K. Sasagane, and K. Yamaguchi, *Int. J. Quantum Chem.* **65**, 665-677 (1997).
7. M. Shiga, F. Aiga, and K. Sasagane, *Int. J. Quantum Chem.* **71**, 251-271 (1999).
8. T. Kobayashi, K. Sasagane, F. Aiga, and K. Yamaguchi, *J. Chem. Phys.* **110**, 11720-11733 (1999).
9. T. Kobayashi, K. Sasagane, F. Aiga, and K. Yamaguchi, *J. Chem. Phys.* **111**, 842-848 (1999).
10. T. Kobayashi, K. Sasagane, and K. Yamaguchi, *J. Chem. Phys.* **112**, 7903-7918 (2000).
11. T. Kobayashi, F. Aiga, K. Sasagane, and K. Yamaguchi, *unpublished*.
12. K. Sasagane, T. Kobayashi, M. Shiga, F. Aiga, and K. Yamaguchi, *Nonlinear Optics* **26**, 33-42 (2000).
13. K. Mori and C. Ohe, *Mem. Kokushikan Univ. CIS* **18**, 62-100 (1997).
14. K. Taneda and K. Mori, *Chem. Phys. Letters* **298**, 293-301 (1998).
15. K. Mori, *Mem. Kokushikan Univ. CIS* **20**, 50-67 (1997).
16. R. Klingbeil, V. G. Kaveeshwar, and R. P. Hurst, *Phys. Rev. A* **4**, 1760-1767 (1971).
17. R. S. Watts and A. T. Stelbovics, *Chem. Phys. Letters* **61**, 351-354 (1979).
18. D. P. Santry and T. E. Raidy, *Chem. Phys. Letters* **61**, 413-416 (1979).
19. D. P. Santry and T. E. Raidy, *Theoret. Chim. Acta* **53**, 121-128 (1979).
20. H. Sekino and R. J. Bartlett, *J. Chem. Phys.* **85**, 976-989 (1986).
21. F. Aiga, K. Sasagane, and R. Itoh, *Chem. Phys.* **167**, 277-290 (1992).
22. S. P. Karna, *J. Chem. Phys.* **104**, 6590-6605 (1996).
23. P. Albersten, P. Jørgensen, and D. L. Yeager, *Mol. Phys.* **41**, 409-420 (1980).
24. D. Yeager, J. Olsen, and P. Jørgensen, *Int. J. Quantum Chem. Symp.* **15**, 151 (1981).
25. H. Hettema and P. E. S. Wormer, *J. Chem. Phys.* **93**, 3389-3396 (1990).
26. M. Jaszuński and R. McWeeny, *Mol. Phys.* **46**, 863-873 (1982).
27. K. Sasagane, K. Mori, A. Ichihara, and R. Itoh, *J. Chem. Phys.* **92**, 3619-3632 (1990).
28. H. J. Aa. Jensen, P. Jørgensen, H. Hettema, and J. Olsen, *Chem. Phys. Letters* **187**, 387-390 (1991).
29. H. Hettema, H. J. Aa. Jensen, P. Jørgensen, and J. Olsen, *J. Chem. Phys.* **97**, 1174-1190 (1992).
30. M. Jaszuński, P. Jørgensen, and H. J. Aa. Jensen, *Chem. Phys. Letters* **191**, 293-298 (1992).
31. D. Jonsson, P. Norman, and H. Ågren, *J. Chem. Phys.* **105**, 6401-6419 (1996).
32. R. Kobayashi, H. Koch and P. Jørgensen, *Chem. Phys. Letters* **219**, 30-35 (1994).
33. C. Hättig, O. Christiansen, H. Koch, and P. Jørgensen, *Chem. Phys. Letters* **269**, 428-434 (1997).
34. C. Hättig, O. Christiansen, and P. Jørgensen, *Chem. Phys. Letters* **282**, 139-146 (1998).
35. C. Hättig and P. Jørgensen, *J. Chem. Phys.* **109**, 2762-2778 (1998).

36. O. Christiansen, S. Coriani, J. Gauss, C. Hättig, P. Jørgensen, F. Pawłowski, and A. Rizzo, "Accurate Nonlinear Optical Properties for Small Molecules. Methods and Results," in *Non-Linear Optical Properties of Matter: From Molecules to Condensed Phases*, edited by M. G. Papadopoulos, A. J. Sadlej, and J. Leszczynski, Dordrecht: Springer, 2006, pp. 51-99.

37. P. Salek, T. Helgaker, T. Saue, *Chem. Phys.* **311**, 187-201 (2005).

38. A. Ye and J. Autschbacha, *J. Chem. Phys.* **125**, 234101 (2006).

39. B. Jansik, P. Sałek, D. Jonsson, O. Vahtras, H. Ågren, *J. Chem. Phys.* **122**, 054107 (2005).

40. D. Jonsson, O. Vahtras, B. Jansik, Z. Rinkevicius, P. Sałek, and H. Ågren, "Kohn–Sham Time-Dependent Density Functional Theory with Applications to Linear and Nonlinear Properties," in *Non-Linear Optical Properties of Matter: From Molecules to Condensed Phases*, edited by M. G. Papadopoulos, A. J. Sadlej, and J. Leszczynski, Dordrecht: Springer, 2006, pp. 151-209.

Chemical Events in Solution Phase. The View from Molecular Level Descriptions

Hirofumi Sato, Kenji Iida, Daisuke Yokogawa, Atsushi Ikeda and Shigeyoshi Sakaki

Department of Molecular Engineering, Kyoto University, 615-8510, Kyoto, Japan

Abstract. Two recently developed theoretical tools to deal with chemical processes in solution phase are presented. One is RISM-SCF theory that is a hybrid method of quantum chemistry and statistical mechanics for molecular liquids. The other is an new analysis method for wave function in terms of resonance structure based on the second quantization technique.

Keywords: RISM-SCF, electronic structure, solvation structure, chemical reaction, resonance structure
PACS: 31.10.+z, 31.15.Ar, 31.15.Bs, 31.15.Ne, 31.70.Dk, 34.50.Lf,61.20.Gy,61.25.Em

INTRODUCTION

The electronic structure of a molecule is significantly changed when the molecule is dissolved into solvent. It means electronic structure, which is treated by quantum chemistry, and solvation structure, which is governed by statistical mechanics, are coupled each other. In this regards, development of new theories that can bridge between these two methodologies is very important to understand ordinary chemical processes.

In the last few decades, the theory of the electronic structure of solvated molecules has attracted many researchers' attentions, and various types of combination methods have been developed. These can be roughly classified into three categories, namely, the dielectric continuum model such as polarizable continuum model (PCM) [1], quantum mechanics/molecular mechanics (QM/MM), and method based on the integral equation theory for liquids.

We have been developing RISM-SCF [2, 3], which is a representative of the third category. The method compiles two ab initio methods in theoretical chemistry: one is the reference interaction site model (RISM) [4, 5], and the other is ab initio molecular orbital (MO) theory. The method determines the electronic structure and the statistical solvent distribution around the solute in self-consistent manner. It is a remarkable advantage that the RISM-SCF method can provide information on the microscopic solvation structure based on the statistical mechanics. The method has been successfully applied to numerous molecular phenomena including chemical reactions, chemical equilibria, charge transfer processes and so on [6]. The three-dimensional version of the method (3D-RISM-SCF) is also developed [7].

Meanwhile the change in the electronic structure of a molecule due to the solvent effect leads to modification on various molecular properties. The questions arising here are how the solvation changes the nature of chemical bond, namely, whether a chemical bond is strengthened by solvation. The concept of the resonance structure is simple but can be a useful tool to understand the electronic structure of a molecule; a bond in which bond-making electrons are remarkably localized on one of the two atoms is called "ionic", while a bond in which the electrons are shared by the two atoms is called "covalent". The electronegativity could offer a heuristic explanation of such bonding, while nowadays, it is easy for the modern quantum-chemical technique to compute bond energy or detailed information on bonding wave functions. However, it is difficult to directly bridge between these traditional concepts and the results of modern computations since the wave function of a molecule is usually described in terms of MOs. To understand chemical processes in solution phase efficiently and clearly, development of new analysis methods from the view point of such considerations is highly desired.

In this contribution, we will review the RISM-SCF method together with its recent applications and improvement. The new resonance structure theory that bridges between the traditional concepts and the results of modern computations will be also presented.

CP1046, *Selected Papers from ICNAAM 2007 and ICCMSE 2007*
edited by T. E. Simos, G. Maroulis, G. Psihoyios, and Ch. Tsitouras
© 2008 American Institute of Physics 978-0-7354-0574-5/08/$23.00

RISM-SCF THEORY

The RISM-SCF method compiles *ab initio* electronic structure theory and statistical mechanical theory of molecular liquid. Total energy of the solvation system is defined as the sum of quantum chemical energy of the solute (E_{solute}) and solvation free energy ($\Delta\mu$) [3].

$$\mathscr{A} = E_{\text{solute}} + \Delta\mu = \langle \Psi | H_0 | \Psi \rangle + \Delta\mu, \tag{1}$$

Since the electronic structure of the solute molecule and the solvation structure around it are determined in a self-consistent manner, the wave function of the solute molecule is distorted from that in isolated state. The energy difference between the solute in isolated state (E_{isolated}) and that in solution phase (E_{solute}) is a quantity to measure the contribution of "solvation effects" to the electronic structure.

$$\begin{aligned} E_{\text{reorg}} &= E_{\text{solute}} - E_{\text{isolated}} \\ &= \langle \Psi | H | \Psi \rangle - \langle \Psi_0 | H | \Psi_0 \rangle, \end{aligned} \tag{2}$$

where $|\Psi\rangle$ and $|\Psi_0\rangle$ are wave functions in solution and in gas phase, respectively. Solvation free energy in the present framework of the theory (excess chemical potential derived with hyper-netted chain approximation) is given by,

$$\Delta\mu = -\frac{\rho}{\beta} \sum_{\alpha s} \int d\mathbf{r} \left[c_{\alpha s}(r) - \frac{1}{2} h_{\alpha s}^2(r) + \frac{1}{2} h_{\alpha s}(r) c_{\alpha s}(r) \right], \tag{3}$$

h and c are the total and direct correlation functions, respectively. $\beta = 1/k_B T$, where k_B and T are the Boltzmann constant and temperature, respectively.

Applying variational principle to Eq. (1), the Fock operator of the RISM-SCF theory (F^{solv}) including a solute-solvent interaction, V, is naturally derived.

$$F^{\text{solv}} = F^{\text{gas}} + V. \tag{4}$$

The interaction energy between the solute and solvent reaction field is given by,

$$\langle \Psi | V | \Psi \rangle = \sum_{\alpha} V_{\alpha} \langle \Psi | b_{\alpha} | \Psi \rangle = \sum_{\alpha} V_{\alpha} q_{\alpha}, \tag{5}$$

where b_{α} is the proper population operator for atom α in the solute molecule and V_{α} is the electrostatic potential acting on this atom.

$$V_{\alpha} = \rho \sum_{s} q_s \int \frac{g_{\alpha s}(r)}{r} d\mathbf{r}. \tag{6}$$

Here $g_{\alpha s}(r) (\equiv h_{\alpha s}(r) - 1)$ is the pair correlation function (PCF) around the solute molecule, namely radial distribution function. Note that V_{α} in the present theory is computed from the microscopic information on the solvation structure.

Since the Fock operator is modified, this is an extension of standard ab initio MO theory that can handle the solvent effect. And if we replace E_{solute} in Eq. (1) with MM energy or a constant value, it is equivalent to the original RISM equation. From this standpoint, the RISM-SCF theory can be regarded as an extension of RISM theory as well as that of ab initio molecular orbital theory.

Very recently, we developed the new-generation of RISM-SCF, RISM-SCF-SEDD [8]. In this method, the auxiliary basis sets (ABSs) on each atom are prepared to divide electron density into the components assigned on each atom. Gill, Johnson, Pople, and Taylor proposed a procedure to determine ABSs which reproduce the electrostatic potential (ESP) provided by MO calculation (GJPT procedure) [9]. The great advantage of GJPT procedure is that it treats directly spatial electron density distribution (SEDD) and does not require the set of grid points that was necessary to fit ESP; it is free from these artificial parameters. Furthermore the RISM-SCF-SEDD is much more robust in the connection between RISM and MO calculation than the original version of RISM-SCF and significantly expands the versatility of the RISM-SCF family [10].

THE NEW RESONANCE STRUCTURE THEORY

By using this procedure, the weights of resonance structures can be computed from MOs utilizing localization of MOs and second quantization. Since the details of the theory has been already published elsewhere [11], just a brief outline is provided here.

In the consequence of the theory, the weights (probabilities) of resonance structure in a (localized) valence orbital ϕ_i between atoms A and B are expressed as follows ($\sigma_1 \neq \sigma_2$),

$$A^- B^+ \quad : \quad \sum_{\mu \in A} \sum_{\nu \in A} \langle \phi_i | \chi_\nu^{\sigma_1+} \chi_\mu^{\sigma_2+} \varphi_\nu^{\sigma_2-} \varphi_\mu^{\sigma_1-} | \phi_i \rangle, \tag{7}$$

$$A - B \quad : \quad 2 \sum_{\mu \in A} \sum_{\nu \in B} \langle \phi_i | \chi_\nu^{\sigma_1+} \chi_\mu^{\sigma_2+} \varphi_\nu^{\sigma_2-} \varphi_\mu^{\sigma_1-} | \phi_i \rangle, \tag{8}$$

$$A^+ B^- \quad : \quad \sum_{\mu \in B} \sum_{\nu \in B} \langle \phi_i | \chi_\nu^{\sigma_1+} \chi_\mu^{\sigma_2+} \varphi_\nu^{\sigma_2-} \varphi_\mu^{\sigma_1-} | \phi_i \rangle, \tag{9}$$

where χ_μ^+ and φ_ν^- are, respectively, creation and annihilation operators related to biorthogonal atomic orbital basis functions, μ and ν. Since there is another contribution from essentially three (or higher)-body correlations, which are usually negligible, sum of these four components is exactly normalized to 1. The weights of resonance structures of a whole molecule can be obtained by multiplication of the weights of the valence LMOs ($w^i(A^- B^+), w^i(A - B), w^i(A^+ B^-)$ and $w^i(\text{other})$).

$$1 = \prod_i^{\text{LMOs}} \left\{ w^i(A^- B^+) + w^i(A - B) + w^i(A^+ B^-) + w^i(\text{other}) \right\}, \tag{10}$$

where $w^i(\text{other})$ corresponds to the three (or higher)-body correlations. The choice of LMOs is related to the choice of valence bond configuration. One of the big advantages of the method is that the weight is easily computed from the one-body density matrix without extra costs.

Table 1 illustrates the weights of representative resonance structures both in gas and in the aqueous solution phase. The electronic structure in the latter case is solved with RISM-SCF theory. In both phases, the most important contribution arises from structure **2**, in which one of the O–H bonds is polarized and the other forms covalent bonding. The contributions from the completely covalent character (**1**) and from the doubly ionized character (**4**) are comparable.

Speculating from its auto-ionization character, the bond polarization of the O–H should be enhanced in aqueous solution. Namely, the weights of the resonance structure including these characters must increase upon transfer from the gas phase to the condensed phase. As expected, the weights associated with ionic structures tend to increase in an aqueous solution. The importance of **2** is furthermore enhanced, and the contribution from **4** exceeds that from **1**. In summary, the weights related to ionic character tend to be grater. On the other hand, the weight of covalent character decreases in an aqueous solution. The contribution from 'other' terms is a negative value. Note that the Mulliken population and Mayer's covalent bond order index sometimes give negative value. Since the second quantized operators used in the present study are closely related to the Mulliken population analysis, the negative value that appeared here is likely to have the same origin. In our experience, the Löwdin type operator gives no negative value of weight [12]. The weights computed by the TZP basis sets are also shown in Table 1. As seen in the table, basis-set dependence is negligible while the Mulliken populations on the oxygen atom show greater dependence. In our experience, the basis-set dependence is always negligible in the present analysis [12].

SUMMARY

The solvation effect must modify the character of a chemical bond in the solute molecule, and, at the same time, this modification in the electronic structure affects the structure of the solvent molecules surrounding the solute. In this contribution, several powerful tools to investigate chemical processes in solution phase were presented. The recently developed RISM-SCF-SEDD significantly expands the versatility of the RISM-SCF family, which can provide both of electronic structure and solvation structure. The aim of the new resonance theory is to characterize chemical bond in condensed phase, which is definitely one of the central subjects in physical chemistry.

TABLE 1. Weights of resonance structures in H_2O

No.		DZP		TZP	
		gas	aq.	gas	aq.
1	O, H H	0.207	0.179	0.204	0.173
2	O^-, H H^+	0.401	0.419	0.403	0.422
3	O^+, H H^-	0.106	0.076	0.104	0.071
4	O^{2-}, H^+ H^+	0.194	0.246	0.199	0.258
5	$^+O^-$, H^- H^+	0.103	0.089	0.103	0.086
6	O^{2+}, H^- H^-	0.014	0.008	0.013	0.007
	other	-0.025	-0.017	-0.026	-0.017
	total	1.000	1.000	1.000	1.000
Mulliken Population (O)		8.6605	8.7967	8.6826	8.8351

ACKNOWLEDGMENTS

We acknowledge financial support by the Grant-in Aid for Scientific Research on Priority Areas "Water and biomoleculesＡh (430-18031019), and by the Grant-in Aid for Encouragement of Young Scientists (17750012), both from the Ministry of Education, Culture, Sports, Science and Technology (MEXT) Japan.

REFERENCES

1. S. Miertus, E. Scrocco, and J. Tomasi, *Chem. Phys.*, **55**, 117 (1981).
2. S. Ten-no, F. Hirata, and S. Kato, *Chem. Phys. Lett.*, **225**, 202 (1994).
3. H. Sato, F. Hirata, and S. Kato, *J. Chem. Phys.*, **105**,1546 (1996).
4. D. Chandler, H. C. Andersen, *J. Chem. Phys.*, **57**, 1930 (1972).
5. F. Hirata, P. J. Rossky, *Chem. Phys. Lett.*, **83**, 329 (1981).
6. *Molecular Theory of Solvation*, edited by F. Hirata, Kluwer Academic Publishers, Dordrecht Hardbound, 2003.
7. H. Sato, A. Kovalenko and F. Hirata, *J. Chem. Phys.*, **112**, 9463 (2000).,
8. D. Yokogawa, H. Sato and S. Sakaki, *J. Chem. Phys.*, in press.
9. P. M. W. Gill, B. G. Johnson, J. A. Pople, and S. W. Taylor, *J. Chem. Phys.*, **96**, 7178 (1992).
10. K. Iida, D. Yokogawa, H. Sato and S. Sakaki, *Chem. Phys. Lett.*, submitted.
11. A. Ikeda, Y. Nakao, H. Sato and S. Sakaki, *J. Phys. Chem.*, A, **110**,9028 (2006); A. Ikeda, D. Yokogawa, H. Sato and S. Sakaki, *Chem. Phys. Lett.*, **424**,449 (2006).
12. A. Ikeda, Y. Nakao, H. Sato and S. Sakaki, to be submitted.

Quantal Cumulant Dynamics for Dissipative Systems

Yasuteru Shigeta

Department of Physics, Graduate School of Pure and Applied Science, Tsukuba University, Tennodai 1-1-1, Ibaraki 305-8571, Japan.

Abstract. We develop a quantal cumulant dynamics method for the quantum tunneling in dissipative environment. Reduced equations of motion of classical and quantal cumulant variables without bath degrees of freedom are derived. We observed suppression of the tunneling that depends on the sign of a friction constant for an Ohmic approximation and on the magnitude of a bath frequency for a single bath mode approximation. A possible mechanism of the suppression is explored by analyzing an effective quantal potential of the tunneling path.

Keywords: dissipative tunneling
PACS: 03.65.Ge,03.65.Sq,03.65.-w,03.65.Xp

INTRODUCTION

The motion of a quantum particle interacting with harmonic bath modes is of considerable interest, with a broad range of applications including tunneling in biological long-range electron and short-range proton transfer reactions. Nevertheless, the many-body quantum simulation is limited to some degree, because it costs too much. Path integral techniques developed by Feynman [1] and Caldeira-Leggett [2] allow a rigorous formal elimination of the bath degrees of freedom, and result in a reduced description for the particle coordinates alone. However, evaluating the necessary path integrals using the nonlocal action in the phase space remains a complex task. Therefore it is desirable to establish an efficient scheme in order to understand the many-body quantum tunneling in dissipative systems.

Instead of the full quantum dynamics, a wide variety of methods based upon semi-classical and quantal dynamics has been proposed. Recently, Prezhdo and co-workers developed the quantized Hamilton dynamics (QHD) approach derived from a different route, i.e. Heisenberg's equations of motion [3, 4, 5, 6, 7, 8]. We have applied the QHD method to the molecular vibrational analysis, and found that the method gives less error in the vibrational frequency that the classical dynamics does [9]. Nevertheless, the QHD in a general potential has not been done yet, because of its tedious derivation. Recently we have developed a new methodology for one-dimensional quantum systems called quantal cumulant dynamics (QCD) as an alternative to the QHD, where coordinates, momenta, and cumulants are central variables [10, 11]. The key ideas in this work are that a shift operator acting on an arbitrary operator is introduced and a cumulant expansion is applied to evaluate the expectation value of the shift operator. The expectation value of the shift operator is independent of the Hamiltonian so that the scheme is applicable to the general potential.

In this paper, we derive reduced equations of motion of cumulants for one-dimensional cases in order to investigate tunneling motion in dissipative systems based on the QCD method. An extension to the multi-dimensional cases is straightforward.

THEORY

The Hamiltonian of a one-dimensional system coupled with classical bath degrees of freedom is given by

$$\hat{H} = \frac{\hat{p}^2}{2M} + V(\hat{q}) + \sum_j \left(\frac{p_j^2}{2m_j} + \frac{m\omega_j^2 x_j^2}{2} + c_j x_j \hat{q} \right) , \qquad (1)$$

where the first term denotes the kinetic energy operator, the second term is the potential energy operator, and third term is Hamiltonian of the classical bath and coupling between the classical bath and the quantum particle. The ordinary commutation relation for $\hat{q}$ and $\hat{p}$ holds, $[\hat{q}, \hat{p}] = i$ $(\hbar = 1)$.

CP1046, *Selected Papers from ICNAAM 2007 and ICCMSE 2007*
edited by T. E. Simos, G. Maroulis, G. Psihoyios, and Ch. Tsitouras

According to the previous works [10, 11], the potential energy can be approximated in terms of the expectation value of the position operator, $q = \langle \hat{q} \rangle$, and the second-order position cumulant variable, $\lambda_{2,0}$, as [12, 13]

$$\langle V(q) \rangle_2 = \exp\left[\sum \frac{\lambda_{2,0}}{2} \frac{\partial^2}{\partial q^2}\right] V(q) . \tag{2}$$

Note here that there are two other cumulants $\lambda_{0,2}$ and $\lambda_{1,1}$ within second-order approximation, where the subscript in $\lambda_{m,n}$ labels mth- and nth-order with respect to the position and momentum, respectively. The second-order total energy is given by

$$E_2 = \frac{p^2 + \lambda_{0,2}}{2M} + \exp\left[\frac{\lambda_{2,0}\nabla^2}{2}\right] V(q) + \sum_j \left(\frac{p_j^2}{2m_j} + \frac{m_j\omega_j^2 x_j^2}{2} + c_j x_j q\right) , \tag{3}$$

where p is the expectation value of the momentum operator, i.e. $p = \langle \hat{p} \rangle$. Hereafter, we refer to the second term as a second-order quantal potential, that is abbreviated as

$$\tilde{V}(q, \lambda_{2,0}) \equiv \exp\left[\frac{\lambda_{2,0}\nabla^2}{2}\right] V(q) = \frac{1}{\sqrt{2\pi\lambda_{2,0}}} \int ds \exp\left[-\frac{(q-s)^2}{2\lambda_{2,0}}\right] V(s) . \tag{4}$$

Note here that if all cumulant variables in the total energy are zero, $\lambda_{2,0} = \lambda_{0,2} = 0$, the energy is identical with that which appears in the classical mechanics.

We next consider the Heisenberg equations of motion (EOM) for the cumulants up to second-order. Since all the classical bath degrees of freedom are treated as harmonic oscillators with linear couplings, one can eliminate the classical bath degrees from the EOM [14]. In the second-order cumulant approach, we result in the reduced EOM of only five variables given by

$$\dot{q}(t) = \frac{p(t)}{M} , \tag{5}$$

$$\dot{p}(t) = -\tilde{V}^{(1)}(q(t), \lambda_{2,0}(t)) + \int_0^t K(t-t')q(t')dt , \tag{6}$$

$$\dot{\lambda}_{2,0}(t) = \frac{2\lambda_{1,1}(t)}{M} , \tag{7}$$

$$\dot{\lambda}_{0,2}(t) = -2\lambda_{1,1}(t)\tilde{V}^{(2)}(q(t), \lambda_{2,0}(t)) , \tag{8}$$

$$\dot{\lambda}_{1,1}(t) = -\lambda_{2,0}(t)\tilde{V}^{(2)}(q(t), \lambda_{2,0}(t)) + \frac{\lambda_{0,2}(t)}{M} , \tag{9}$$

$$\tag{10}$$

where $\tilde{V}^{(n)}(q(t), \lambda_{2,0}(t))$ is the n-th derivative of the quantal potential with respect to $q(t)$. It should be stressed here that the term arising from the classical bath appears only in the second line and is nonlocal in time so that it describes a friction with memory effects. The memory kernel, $K(t-t')$, is represented as

$$K(t-t') = \sum_j \frac{c_j^2}{m_j\omega_j} \sin\left[\omega_j (t-t')\right] . \tag{11}$$

The above equation indicates that if one has all the coupling coefficients and the harmonic frequencies, then the memory kernel can be evaluated explicitly. In the so-called Ohmic approximation, $K(t-t') = \gamma(t)\delta(t-t')$, the friction term is assumed to be local and a simple function of $q(t)$ multiplied by $\gamma(t)$, where $\gamma(t)$ is an effective friction coefficient.

Here we describe initial conditions in the actual numerical simulation. Due to the Heisenberg's uncertainty relation and a variational principle of the energy, the initial conditions of the variables can be determined. In particular, $\lambda_{2,0}\lambda_{0,2} = 1/4$ and $\lambda_{1,1} = 0$ that ensures the least quantal energy principle initially suggested by Tsue [15].

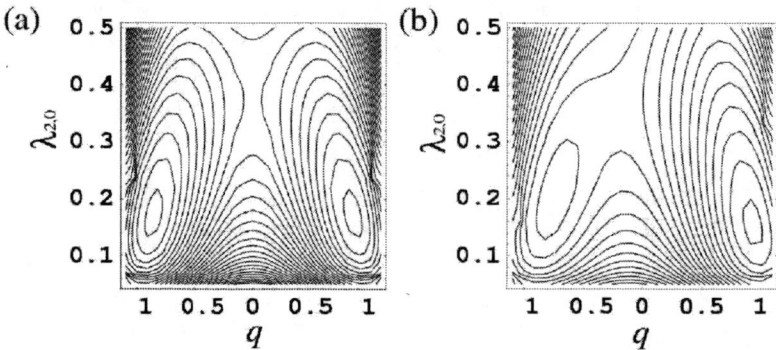

FIGURE 1. Effective quantal potential (a) without and (b) with a linear coupling, $c(t) = -0.4$, respectively.

NUMERICAL DETAIL AND RESULTS

We treat a double well potential given by

$$V_W(q) = q^4 - 2q^2 + c(t)q, \tag{12}$$

where the last term is a linear coupling due to the classical bath and $c(t) = \sum_j c_j x_j(t)$ is an effective linear coupling coefficient. The tunneling problems with a dissipative coupling may be regarded as the QCD when the particle is initially located at one of the well, and then tunnels through the barrier to the other well. Before we proceed to the dynamics simulation, we analyze how the tunneling motion is described in the present method.

By substituting the potential in Eq. (4), we have

$$\tilde{V}_W(q, \lambda_{2,0}) = q^4 - 2q^2 + c(t)q + \lambda_{2,0}(3q^2 - 2) + 3\lambda_{2,0}^2. \tag{13}$$

We here define the effective potential that includes quantal correction to the kinetic energy operator as

$$\tilde{V}_{eff}(q, \lambda_{2,0}) = \frac{1}{8M\lambda_{2,0}} + \tilde{V}_W(q, \lambda_{2,0}), \tag{14}$$

where the first term originates from the least quantal energy principle, i.e. $\lambda_{2,0}\lambda_{0,2} = 1/4$. The effective potential of the double well for $c(t) = 0$ is plotted in Fig. 1-(a) with $M=3.5$. It is found that there are two minima at $(q, \lambda_{2,0}) = (\pm 0.80486, 0.11740)$, which is a global minimum of the effective potential, and the energy is $E_{min} = -0.30889$. The transition state of the effective potential is at $(q, \lambda_{2,0}) = (0, 0.37554)$ and the corresponding energy is $E_{eff}^{TS} = -0.23289$. Since the corresponding classical transition state is $E_c^{TS} = 0$ at $q = 0$, the transition state reflects quantum nature of the tunneling motion.

In the dissipative system, the effective potential is time dependent, then the transition state moves and the barrier varies with time. For example, if the linear coupling term with $c(t) = -0.4$ is added, the potential becomes non-symmetric and one of two minima becomes more stable than the other (see Fig. 1-(b)). The transition state energy is $E_{eff}^{TS} = -0.18588$ and the energies of the global and local minima are $E_g = -0.95928$ and $E_{loc} = -0.27255$, respectively. This indicates two different mechanisms as follows: (i) if the particle is initially located around the right minimum, this state is stabilized due to the friction term, and then the tunneling is suppressed because of the higher barrier and (ii) if the particle is initially located around the left minimum, a transition from left to right minimum may undergo because of the shallower barrier. Later we observe the tendency in the QCD simulation.

For the QCD simulation, the initial total energy is estimated to be $E(0) = -0.21896$ at $(q, \lambda_{2,0}) = (1, 0.08877)$, on the other hand, the energy of quantal transition state of the effective potential is $E_{eff}^{TS} = -0.23289$. It should be stressed here that the total energy is slightly higher than that of the quantal transition state so that the quantum tunneling occurs. The time interval is set at $\delta t = 0.01$ (a.u.) and total time is $T = 40$ for the Ohmic approximation with a time-independent friction constant, $\gamma(t) = \text{const.}$, and $T = 400$ for a single bath mode approximation with $c_1^2/\omega_1 m_1 = 0.01$. The fourth-order Runge-Kutta integrator is used for the time integration.

Figures 2 depict phase space structures ($q - p$ plane) by (a) the Ohmic and (b) the single bath mode approximations. In Fig. 2-(a), tunneling trajectories are found for $\gamma = 0$ and $\gamma = 0.01$. The fact indicates that the quantum tunneling

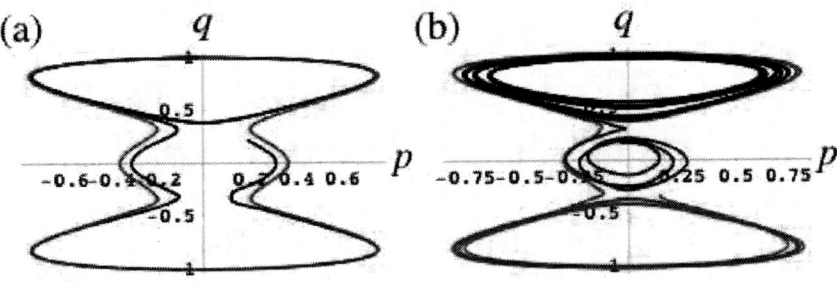

FIGURE 2. Tunneling trajectory by (a) Ohmic approximation (red for $\gamma = 0.01$, blue for $\gamma = -0.01$, and black for $\gamma = 0$), and (b) with a single bath mode with $c_1^2/\omega_1 m_1 = 0.01$ (red for $\omega_1 = 0.2$, blue for $\omega_1 = 0.4$, and black for $\omega_1 = 0.6$).

occurs both without the friction and with the positive friction term. On the other hand, it does not occur for the negative friction term. The behavior corresponds to the mechanism of the suppression analyzed above. In Fig. 2-(b), tunneling trajectories are also found, where the particle transits from one well to the other intermittently for $\omega_1 = 0.2$ and is stagnant around $q = 0$ rather than the minima during the tunneling for $\omega_1 = 0.4$. These behaviors are quite different from that under the Ohmic approximation. As the ω_1 value increases, tunneling motion is suppressed. In particular, the particle is confined in one well like as in the Ohmic case with the negative friction coefficient. Applications of multi-dimensional and multi-mode cases are now in progress.

CONCLUSION

We have formulated a quantal cumulant dynamics method for the particle interacting with a linear classical bath. In particular, we have derived reduced equations of motion for the position, momentum, and second-order cumulants variables. We have numerically shown (1) mechanisms of the suppresion of tunneling by the tunneling transition states and energy minima on the effective quantal potential and (2) the suppression of the tunneling due to the classical bath with the Ohmic and single bath approximations.

ACKNOWLEDGMENTS

This research was supported by the Core Research for Evolutional Science and Technology (CREST) Program "High Performance Computing for Multi-Scale and Multi-Physics Phenomena" of the Japan Science and Technology Agency.

REFERENCES

1. R.P. Feynman, F.L. Vernon, Ann. Phys. **24**, 118 (1963).
2. A.O. Caldeira, A.J. Leggett, Ann. Phys. **149**, 374 (1982).
3. O.V. Prezhdo and Y.V. Pereverzev, J. Chem. Phys. **113** 6557 (2000); ibid **116**, 4450 (2002).
4. E. Pahl and O.V. Prezhdo, J. Chem. Phys. **116**, 8704 (2002).
5. O.V. Prezhdo, J. Chem. Phys. **117** 2905 (2002).
6. D. S. Kilin, Y. V. Pereversev, O. V. Prezhdo J. Chem. Phys. **120**, 11209 (2004).
7. E. Heatwole and O.V. Prezhdo, J. Chem. Phys. **121**, 10967 (2004); ibid **122**, 234109 (2005).
8. O.V. Prezhdo, Theoret. Chem. Acc. **116**, 206 (2006).
9. H. Miyachi, Y. Shigeta, K. Hirao, Chem. Phys. Lett. **432**, 585 (2006).
10. Y. Shigeta, H. Miyachi, K. Hirao, J. Chem. Phys. **124**, 244102 (2006).
11. Y. Shigeta, H. Miyachi, K. Hirao, Chem. Phys. Lett. in press.
12. J. E. Mayer, J. Chem. Phys. **5**, 67 (1937).
13. R. Kubo, J. Phys. Soc. Jpn. **17**, 1100 (1962).
14. M. Razavy, Phys. Rev. **A41**, 6668 (1990).
15. Y. Tsue and Y. Fujiwara, Prog. Theoret. Phys. **86**, 443 (1991).

Multi-scale Approach to the Design of Ferroelectric Liquid Crystals for NLO Applications

Armand Soldera[†], Frédéric Perreault[†] and Benoît Champagne[††]

[†]Centre d'Études des Matériaux Optiques et Photoniques de l'Université de Sherbrooke, Département de Chimie, Université de Sherbrooke, 2500 Boul. Université, Sherbrooke, Québec, Canada J1K 2R1
[††]Laboratoire de Chimie Théorique Appliquée, Facultés Universitaires Notre-Dame de la Paix, rue de Bruxelles, 61, B-5000, Namur, Belgium

Abstract. Comparing data from *ab initio* calculations on a molecule and experimental measurements in the material is not straightforward since environment and bulk symmetry have to be considered. To design compounds with performing properties, a multi-scale approach is proposed, combining molecular modeling with experimental and theoretical investigations. This procedure is applied to the design of new ferroelectric liquid crystals for NLO applications. New compounds are proposed by calculations with enhanced NLO properties.

Keywords: Smectic C; push-pull molecules; chiral sulfinate smectogen.
PACS: 31.15.Ar, 42.65.An, 42.65.Ky, 42.70.Df

INTRODUCTION

Molecular modeling is becoming an essential asset in a laboratory. By reproducing experimental values, or by giving an insight in the atomistic origin of a specific property, it greatly contributes to the design of highly performing compounds by suitably tuning molecular interactions. However, calculations performed on one molecule could not be directly compared to experimental values since short and long-range interactions, i.e. its local and global environments, and the bulk symmetry of the material have to be taken into account. A multi-scale approach is then used to get over these limitations. Nevertheless experimental and theoretical investigations have to be carried out together with molecular simulation to design new performing materials, as it is illustrated in this text. The compounds are ferroelectric liquid crystals (FLCs), and the property to optimize is the second-order susceptibility, $\chi^{(2)}$, for nonlinear optical (NLO) applications.

Thermotropic liquid crystal phases, or mesophases, are particular states of matter where mesogens (molecules from which liquid crystal phases are obtained) are aligned according to the interactions between their rigid core, giving rise to anisotropic properties, and their inherent motion is due to the presence of aliphatic chains linked to the rigid core. Among the great number of mesophases, one is of great interest for NLO applications: chiral smectic C phase, or Sm C*. Material exhibiting the smectic C phase presents the C_{2h} symmetry in the Schoenfliess notation. In this mesophase, molecules are tilted by an angle θ from the normal to the plane to which they belong to. Inserting an asymmetry in the medium or making the smectogen C molecules chiral yields the C_2 symmetry to the bulk material. According to the Neumann principle[1], such symmetry gives rise to a non-vanishing property along the axis of C_2 symmetry. At the microscopic level, charge separation results in the appearance of a spontaneous polarization in the material. It is the reason why smectic C* phase is usually called FLC. However, other properties can be envisioned. For NLO applications, a great charge transfer is specifically looked for. Accordingly first FLC materials have been considered as potentially good candidates for NLO applications since non-centrosymmetry is inherent to their structure: FLCs belong to the C_2 group of symmetry. In fact, due to the structure of the smectogen C molecules, the major first hyperpolarizability β component lies along the rigid core which is perpendicular to the C_2 axis lying in the smectic plane. As a matter of fact, first FLC materials exhibited very low values of $\chi^{(2)}$. In 1994, two groups have identified molecules where smectogen C molecules (i.e. molecules that possess the chiral smectic C phase in their polymorphism) carrying the major β component along the C_2 axis. Since that time, no real

CP1046, *Selected Papers from ICNAAM 2007 and ICCMSE 2007*
edited by T. E. Simos, G. Maroulis, G. Psihoyios, and Ch. Tsitouras
© 2008 American Institute of Physics 978-0-7354-0574-5/08/$23.00

progress has been done. The purpose of this article is to apply the multi-scale approach to the design of new smectogen C* molecules carrying a NLO-phore in order to get FLCs with high values of $\chi^{(2)}$.

DESIGN

Overall Approach to the Design

Combining molecular modeling with theoretical and experimental considerations through a multi-scale approach to design FLCs for NLO applications is aimed at filling the knowledge gap that still exists between microscopic and macroscopic appreciations of mesogen molecules and liquid crystalline material respectively. In a multi-scale scheme, these two levels are located at the outmost ends of the spectrum: molecules make up the system where simulation is carried out, and FLC material is the outermost system where experimental measurements are performed. Much has to be done in order to allow for a reliable and predictive use of computational tools for molecular design, i.e. going from the molecule to the material. Moreover, only electronic calculations can be run on a molecule to extract the β components (β is actually a tensor). Experimental measurements give information on the properties averaged over all the possible conformations of molecules, and their stacking symmetry. None of these parameters are explicitly taken into account at the *ab initio* level. Arrangement of molecules could not be solved due to the very long range interactions between molecules in the smectic phase. So a molecule is the most basic constitutive unit to consider in this puzzle, and ideally calculations on a single molecule have to reproduce the intrinsic properties of the bulk material. The design becomes tractable by performing symmetry operations, and by considering the alignment and orientation of the mesogens in the material through the use of a model for the molecule in the smectic C phase. Valuable information is thus extracted from the microscopic data. Nevertheless to validate these data, they have to be compared to experimental measurements which will then confirm or invalidate the approach used. The scheme employed for this purpose is presented in the next paragraph.

Design Applied to FLCs for NLO

The approach used to design efficient compounds for NLO applications is based on tuning molecular characteristics of existing molecules. As a matter of fact, a primary molecule has to be considered. This molecule exhibits the smectic C* phase in its polymorphism. It has been synthesized by Soldera et al.[2], and is displayed in Figure 1. The key feature of this molecule is that its chiral center is directly linked, and thus coupled, to the rigid core carrying the NLO-phore. Actually, from a material viewpoint, the C_2 axis passes through the chiral center. Accordingly the NLO-phore which imparts to the material the nonlinear susceptibility, presents the β component directly linked to the center that carries the non-centrosymmetry axis. Such molecules are good candidates as promising performing molecules with high $\chi^{(2)}$ values for the material since the distance between the chiral center and the NLO-phore is greatly reduced. In fact increasing this distance decreases the possibility of getting an important charge transfer along the C_2 axis, due to an increase in the entropy.

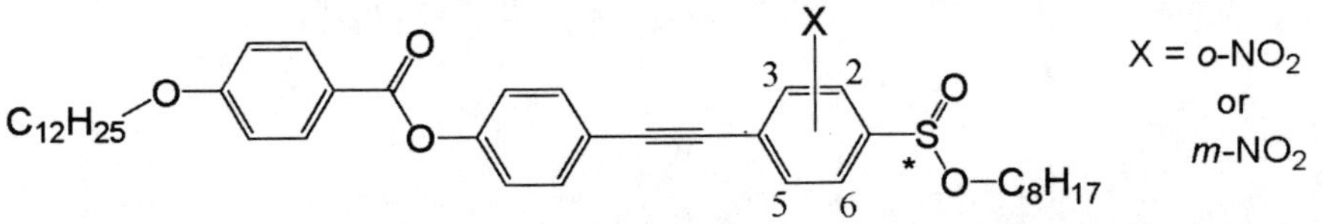

FIGURE 1. Molecule used as a template to generate a family of compounds for nonlinear optical applications.

In fact, smectogen C* molecules such as the one displayed in Figure 1, give helielectric liquid crystals [de Gennes] since the presence of an asymmetry involves the rotation of the projection of the director (pointing along the molecule axis) in the smectic plane (comparable to the rotation of the planes in a cholesteric phase)[3]. To suppress this helix and thus to make the material ferroelectric, a proper surface treatment with a small cell thickness (3-4 μm) is applied during sample preparation[4].

From a practical perspective, the sulfinate phenyl ring of the NLO-phore in molecule of Figure 1 is ortho- or meta- substituted with a nitro electron-acceptor group. Efficient electron density transfer in a π-electron system between a donor-acceptor pair enhances β through the push-pull effect[5]. For instance, paranitroaniline exhibits high first-order hyperpolarizability[5]. Adding an amino electron-donor group on the NLO-phore of the sulfinate compound shown in Figure 1 is thus most likely going to increase β values. However, the positioning of the amino and nitro groups relative to the sulfur atom is crucial. Its influence is actually unveiled by the proposed approach using a combination of theoretical and experimental studies. The hyperpolarizabilities of a family of compounds generated by permuting the amino and/or nitro groups around the phenyl are thus obtained from *ab initio*, or semi-empirical, calculations.

Calculations

From the template molecule in Figure 1, nine different molecules were created by permuting donor and/or acceptor groups around the phenyl. The structures were fully optimized at the B3LYP/6-31G* level. The time-dependent Hartree-Fock (TDHF) scheme was employed using the 6-31+G* basis sets. That scheme consists in expanding the matrices of the TDHF equations in Taylor series of the external electric field and solving them order by order[6]. Both optimization and hyperpolarizability calculations were performed using the Gaussian03 package[7]. All 27 components of the third-rank β tensor were evaluated for each molecule. These microscopic data, as mentioned above, can subsequently be treated as to incorporate characteristic structural information obtained on a different scale by experimental techniques. The detailed approach and treatment are described elsewhere[8] and will be summarized here.

The important quantity to consider in our system is the magnitude of the hyperpolarizability tensor along the axis of C_2 symmetry since it is the only direction in which the material is going to sustain a non zero susceptibility $\chi^{(2)}$. As mentioned above, the axis of C_2 symmetry is perpendicular to the rigid core of the molecules as shown in Figure 2. However, the exact relative position between these two is not known since rotation around the long axis of the molecule is possible. Hence, the magnitude of β along the C_2-axis has to be evaluated for a complete rotation around the long axis of the molecule. The dipole moment induced when the molecule is submitted to an electric field along C_2 by considering the first hyperpolarizability contribution only was calculated. Since the electric field is chosen perpendicular to the rigid core and consists of a unit vector, this operation is simply the evaluation of the tensor component in the C_2 direction, β_{C_2}.

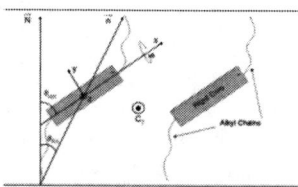

FIGURE 2. Relative position of the axis of C_2 symmetry and the rigid core. β_{C_2} is evaluated for angles Φ between 0 and 360°.

The rigid core is thus rotated around its long axis and β_{C_2} is evaluated every 10°. The data obtained for some of the sulfinate compounds and two reference molecules are presented in Figure 3.

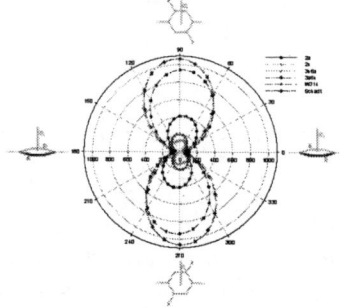

FIGURE 3. Polar plots of the projections of the hyperpolarizability tensors on the two-fold axis of symmetry for a rotation around the long axis of the rigid core.

In the legend of Figure 3, *a* and *n* stand for amino and nitro groups respectively such as the molecule named 3a6n has an amino group in position 3 and a nitro group in position 6 (see Figure 1). The reference molecules named W314 and Schadt are presented elsewhere[9,10]. The Schadt's molecule yields the FLC known in the literature as having the larger $\chi^{(2)}$ among FLCs. Some of our compounds show a larger β_{C_2}. Moreover entropy associated to the C_2 axis and the NLO-phore is strongly reduced. These molecules are thus deserved to get synthesized.

CONCLUSION

The approach exposed in this text brings together *ab initio* calculation in a multi-scale approach to design FLCs for NLO applications. In order to reconcile the two outermost viewpoints a strategy was employed. It is based on merging together experiments and theories in addition to molecular simulation, in order to take into account the huge number of parameters that makes impossible the direct transition from the molecule to the material. The next step is thus to synthesize the proposed molecules. Other difficulties also arise in the measurement of $\chi^{(2)}$. At the final point, revealing the link between NLO properties, microscopic β value and macroscopic $\chi^{(2)}$ value, demonstrates the utmost importance of understanding the molecular arrangement of molecules in the smectic C phase. Moreover such perspective will allow crossing the step to the next level in the multi-scale approach, by specifically using atomistic modeling.

It has to be pointed out that there is one important drawback in the proposed procedure: the molecule that exhibits high β values has to exhibit the ferroelectric phase in its polymorphism. This question can only be solved by synthesizing the molecules, and characterizing them. Moreover, our approach is based on a reductionism viewpoint, i.e. the whole can be reduced to its parts. The emergence perspective indicates that the whole is superior to the sum of its parts. Liquid crystals are typical example of what Laughlin calls protectorate. Clarification of these different standpoints could eventually emerge from the analysis proposed.

ACKNOWLEDGMENTS

The authors are grateful to the CFI, NSERC, CÉMOPUS, the Ministère des Relations Internationales du Québec and the Direction générale des Relations Extérieures de la Région wallonne de Belgique, through the cooperation Québec/Wallonie ST-8/04.808 program. B.C. thanks the Belgian National Fund for Scientific Research for his Research Director position. The calculations have been performed on the Interuniversity Scientific Computing Facility (ISCF), installed at the Facultés Universitaires Notre-Dame de la Paix (Namur, Belgium), for which the authors gratefully acknowledge the financial support of the FNRS-FRFC and the "Loterie Nationale" for the convention n° 2.4578.02, and of the FUNDP.

REFERENCES

1. J.F. Nye, *Physical Properties of Crystals. Their representation by Tensor and Matrices.* (Oxford University Press Inc., New York, New York, 2003).
2. A. Soldera and R. Théberge, *Liquid Crystals* **30** (10), 1251 (2003).
3. P.G. de Gennes and J. Prost, *The Physics of Liquid Crystals.* (Oxford University Press, Oxford, 1993).
4. N.A. Clark and S.T. Lagerwall, *Appl. Phys. Lett.* **36** (1980).
5. J. Zyss, *Nonlinear optical properties of organic molecules and crystals. 1.* (Academic, San Diego, 1987).
6. H. Sekino and R.J. Bartlett, J. Chem. Phys. **85** (1986); S.P. Karna and M. Dupuis, J. Comp. Chem. **12** (1991).
7. M.J. Frisch, G.W. Trucks, H.B. Schlegel et al., GAUSSIAN 03, Revision C.02, Gaussian, Inc., Pittsburgh PA (2003).
8. F. Perreault, B. Champagne, and A. Soldera, *Chemical Physics Letters* **440** (1-3), 116 (2007).
9. D.M. Walba, M.B. Ros, T. Sierra et al., Ferroelectrics **121**, 247 (1991).
10. K. Schmitt, C. Benecke, M. Schadt et al., *J. Phys. III France* **4**, 387 (1994).

Computation of the Reduction Free Energy of Coenzyme in Water: A Novel Approach within the Framework of the QM/MM-ER Method

Hideaki Takahashi,[*] Hajime Ohno, Ryohei Kishi, Shin-ichi Furukawa, and Masayoshi Nakano

Division of Chemical Engineering, Department of Materials Engineering Science, Graduate School of Engineering Science, Osaka University, Toyonaka, Osaka 560-8531, Japan

Abstract. We have computed free energy for one-electron reduction of the active site of the coenzyme (flavin adenine dinucleotide (FAD)) in aqueous solution by means of the QM/MM method combined with the theory of energy representation (QM/MM-ER). In the present work, we have proposed a novel approach that the excess electron to be attached on the FAD has been regarded as a solute, while the remaining molecules including the active site of FAD have been considered as solvent in the implementation of the method of energy representation. The efficiency and the accuracy of the method have been examined by performing the conventional simulations where the oxidized and the reduced active sites of FAD are regarded as solutes and the surrounding water molecules are treated as solvent. The present approach is computationally more advantageous and is amenable to the extension to the reaction in the protein systems as compared with the conventional one. It has been found that the reduction free energy obtained by the present method is in excellent agreement with that computed by the conventional approach, which guarantees the robustness of the method.

Keywords: Free Energy, Real-space Grids, DFT, QM/MM, Theory of Energy Representation, QM/MM-ER
PACS: 31.15.Ew, 31.15.Fx, 71.15.Pd

INTRODUCTION

The pathway of a reaction in condensed phase is dominated by the free energy change associated with the process. Hence, it is a subject of fundamental importance to determine the free energy change in the field of the theoretical and computational studies of chemical processes in condensed phases. In biological systems, in particular, it is possible that the structural fluctuation of the protein regulates the enzymatic reaction and plays a key role in making the protein to exhibit its elaborate function. It should also be noted that the effective motion of the protein for the desired function would be realized by the surrounding water molecules mediating the correlation among the secondary structures that constitutes the protein. Thus, it is essential to incorporate the influence of the dynamics of the protein for the theoretical investigation of the chemical events in biological systems.

One of the major difficulties we encounter to compute the free energy change associated with the reaction in a many-particles system arises from the quantum chemical calculations for large systems. In the *ab initio* electronic structure calculations the computational cost scales in order $N^3 \sim N^4$ with respect to the number N of electrons contained in the system even at low levels of theory. The recent rapid growth in the density functional theory (DFT) enables one to compute the ground state energy as well as the electron density with substantial accuracy within a reasonable computational cost as compared with the conventional wave function theories. However, its application is still limited to relatively small systems. Another situation that complicates the computation of the free energy change is the fact that a considerable amount of ensemble for molecular configurations is needed to attain the convergence in the free energy. Thus, it is a heavily demanding task to compute free energy for a reaction in a condensed system due to the difficulties related to quantum chemistry and the statistical mechanics. In a recent development, Takahashi and Matubayasi proposed a novel approach to compute free energies for solution system by combining the hybrid quantum mechanical/molecular mechanical method and the theory of energy representation (QM/MM-ER) [1]. Within the theory of energy representation developed by Matubayasi[2], the distribution

CP1046, *Selected Papers from ICNAAM 2007 and ICCMSE 2007*
edited by T. E. Simos, G. Maroulis, G. Psihoyios, and Ch. Tsitouras
© 2008 American Institute of Physics 978-0-7354-0574-5/08/$23.00

functions for the solute-solvent interaction potential, instead of spatial distributions, serve as fundamental variables to determine the solvation free energy of a solute. With this approach the free energy change can be computed within a modest computational cost since it requires sampling of the molecular configurations only at the initial and the final states of the chemical process of interest. Most importantly, the diffuseness of the electron density of the solute as well as its fluctuation, which are inherent in the quantum mechanical object, can be naturally incorporated into the free energy calculations. Another notable feature of the method is that it utilizes the real-space grids to express the one-electron wave functions in the Kohn-Sham DFT aiming at the high performance in the parallel computation. The efficiency of the method has been well examined in the previous work[1] which has demonstrated that the solvation free energies can be computed with substantial accuracies.

The electron transfers, i.e. oxidation and reduction reaction, is one of the most common processes among the wide variety of chemical events in biological systems. In the present work, we propose a novel approach to compute reduction free energy of a solute in aqueous solution, which is so developed that it can be readily extended to reduction reaction in protein systems. The point of the method is that only the excess electron to be attached on the solute is regarded as a solute and the remaining molecules including the active site of the system are treated as the solvent in the implementation of the QM/MM-ER. We applies the method to the one-electron reduction of the active site of the coenzyme FAD (Flavin Adenine Dinucleotide) in aqueous solution. The efficiency of the method is examined by making comparisons with the values obtained by the conventional simulations where the oxidized and the reduced molecules are treated as solutes.

METHODS AND COMPUTATIONAL DETAILS

QM/MM-ER Approach

In the QM/MM simulations the Kohn-Sham one-electron wave functions are represented by the values on the real-space grids uniformly distributed over a QM cell. Accordingly, the effective Hamiltonian of the Kohn-Sham equation is also represented by real space. In particular, the kinetic energy operator is approximated by the higher order finite-difference scheme proposed by Chelikowsky et al[3] and the non-local pseudo potentials for the valence electrons are also expressed on the real space. The spin-unrestricted BLYP functional is employed to compute the exchange correlation potential for the electrons. The solvent described by MM force field consists of 676 TIP3P water molecules which are confined within a spherical cavity by a van der Waals wall.

In the theory of energy representation, the solvation free energy $\Delta\mu$ of a solute is expressed in terms of the distribution functions of the solute-solvent interaction energy, thus,

$$\Delta\mu = -k_B T \int d\varepsilon \left[\left(\rho(\varepsilon) - \rho_0(\varepsilon)\right) + \beta\omega(\varepsilon)\rho(\varepsilon) - \beta\left(\int_0^1 d\lambda\omega(\varepsilon,\lambda)\right)\left(\rho(\varepsilon) - \rho_0(\varepsilon)\right) \right] \quad (1)$$

where $\rho(\varepsilon)$ and $\rho_0(\varepsilon)$ are the energy distribution functions for the solution and the reference systems, respectively, and β denotes the reciprocal of $k_B T$. λ in Eq. (1) indicates the coupling parameter introduced to express the gradual solute insertion into solvent and $\omega(\varepsilon, \lambda)$ is the indirect part of the solute-solvent potential of mean force. Integration of $\omega(\varepsilon, \lambda)$ with respect to the coupling parameter is performed by the hybrid functional of the PY and HNC approximations. In the QM/MM-ER approach, the energy distribution functions in Eq. (1) are directly constructed in a series of QM/MM simulations. In the following we describe the outline for the novel and the conventional methods to compute reduction free energy in the framework of QM/MM-ER.

A Novel Approach

The excess electron has been regarded as the solute and the remaining molecules, i.e. the active site of FAD and the water molecules are treated as solvent in the present approach. Thus, the system is regarded as a mixed solvent system. The active site of FAD is called isoalloxazine ring and is schematically depicted in Fig. 1. The isoalloxazine ring is, of course, treated as QM object in the QM/MM simulation. Then, we have to define the interaction energy v_{QM} between the QM object and the excess electron in implementing the theory of energy representation. v_{QM} is defined as

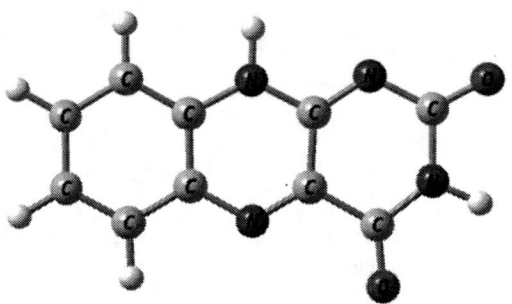

FIGURE 1. The ball and stick model for the isoalloxazine ring in the coenzyme flavin adenine dinucleotide (FAD).

$$v_{QM} = \left\langle \Psi_{N+1} \left| H_0^{QM} \right| \Psi_{N+1} \right\rangle - \left\langle \Psi_N \left| H_0^{QM} \right| \Psi_N \right\rangle \tag{2}$$

where H_0^{QM} is the Hamiltonian for the QM molecule at isolation and $\left| \Psi_{N+1} \right\rangle$ and $\left| \Psi_N \right\rangle$ are the instantaneous electronic eigenfunctions for the reduced and the oxidized isoalloxazine ring, respectively, under the given configuration of solvent water molecules. Then, the energy distribution function is constructed for the interaction potential v_{QM}, which is used as input to Eq. (1) to compute free energy contribution $\Delta\mu_{QM}$ given by the QM molecule to the reduction free energy. On the other hand, the interaction potential v_{MM}^i between the ith water molecule and the excess charge is defined as

$$v_{MM}^i = \sum_m q_m^i \int d\mathbf{r} \frac{n_{exc}(\mathbf{r})}{\left| \mathbf{r} - \mathbf{r}_m^i \right|} \tag{3}$$

where $\mathbf{r}$ is the position vector of the electron and $n_{exc}(\mathbf{r})$ is the density of the excess electron. q_m^i and $\mathbf{r}_m^i$ in Eq. (3) are, respectively, the value of the point charge for mth site of the ith solvent and its coordinate. The distribution functions for the potential v_{MM}^i is also substituted into Eq. (1) to compute free energy contribution $\Delta\mu_{MM}$ due to the interaction between excess charge and the MM water molecules. Then, the total reduction free energy $\Delta\mu_{red}$ can be given by the sum of these contributions, thus, $\Delta\mu_{red} = \Delta\mu_{QM} + \Delta\mu_{MM}$. It should be noted that the method described in this subsection can be readily extended to the reduction of the coenzyme embedded in a protein.

The Conventional Approach

For the purpose to examine the accuracy of the novel approach introduced in subsection 2.2, we have also estimated the free energy $\Delta\mu_{red}$ with the conventional method. By consulting a thermodynamic cycle, it can be readily understood that the reduction free energy $\Delta\mu_{red}$ of the isoalloxazine ring in aqueous solution is expressed in terms of the reduction free energy $\Delta\mu_{red}^{gas}$ in the gas phase and the solvation free energies of the initial and the final states, thus,.

$$\Delta\mu_{red} = \Delta\mu_{red}^{gas} + \left(\Delta\mu_{sol}^{N+1} - \Delta\mu_{sol}^N \right) \tag{4}$$

where $\Delta\mu_{sol}^N$ and $\Delta\mu_{sol}^{N+1}$ are, respectively, the solvation free energies of the oxidized and the reduced isoalloxazine ring. The free energy change $\Delta\mu_{red}^{gas}$ in Eq. (4) can be obtained by performing the frequency analyzes in Gaussian 03 for instance and the solvation free energies $\Delta\mu_{sol}^N$ and $\Delta\mu_{sol}^{N+1}$ are computed by the ordinary QM/MM-ER simulations where the oxidized and the reduced isoalloxazine rings are treated as solute molecules. We refer the readers to the previous paper[1] for the details of computation of the solvation free energies.

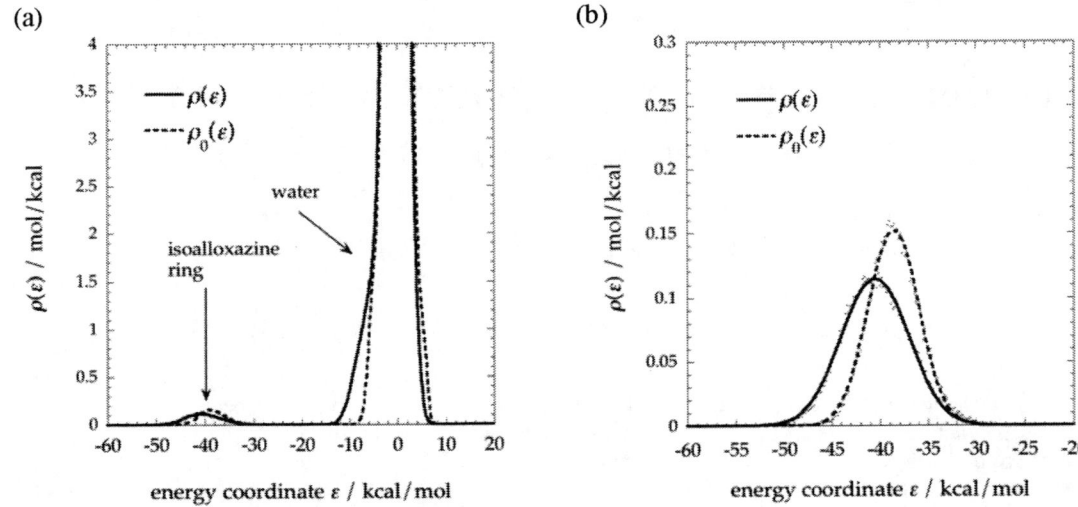

FIGURE 2. (a) Energy distribution functions for the interactions of excess charge with isoalloxazine ring and water molecules. Real and broken lines are for the solution and the reference systems, respectively. (b) Highlighted figure for the distribution for the isoalloxazine ring.

RESULTS AND DISCUSSIONS

Figures 2 show the energy distribution functions for the interaction potentials v_{QM} and v_{MM} defined by Eqs. (2) and (3), respectively. Figure 2(a) depicts the total energy distribution functions for the mixed solvent system and Fig. 2(b) highlights the distributions for the QM molecule. In the distribution function for the interactions between excess charge and the water molecules, a distinguished peak can be seen around the energy coordinate of $\varepsilon = 0$ kcal/mol. However, one finds no prominent structure on the lower energy coordinate corresponding to the strong ion-dipole interactions. The distribution between the excess charge and water molecules lie only in the region of $\varepsilon \gtrsim \sim -12$ kcal/mol. This may be attributed to the fact that the singly occupied molecular orbital (SOMO), on which the excess charge resides, is delocalized over the isoalloxazine ring as a π orbital. The notable feature in the distribution due to the interaction between the excess charge and the isoalloxazine ring treated as QM molecule is that the distribution of the solution system is shifted to lower energy coordinate as compared with that of the reference system. This clearly suggests that the electron affinity of the oxidized QM molecule is enhanced by the hydration. The distribution functions in Fig. 2(a) have been substituted into Eq. (1) to compute reduction free energy $\Delta\mu_{red}$. The free energy contribution $\Delta\mu_{QM}$ from the QM molecule has been computed as −39.5 kcal/mol, while the contribution $\Delta\mu_{MM}$ from water molecules has been estimated as −40.6 kcal/mol including the Born's correction. Totally, we have the reduction free energy $\Delta\mu_{red} = -80.1$ kcal/mol by the present methodology.

As for the conventional approach to the reduction free energy, the solvation free energies for the oxidized and the reduced QM molecules have been obtained as $\Delta\mu_{sol}^{N} = -19.1$ kcal/mol and $\Delta\mu_{sol}^{N+1} = -55.0$ kcal/mol. Then, we have the reduction free energy $\Delta\mu_{red} = -81.0$ kcal/mol by using Eq. (4) where $\Delta\mu_{red}^{gas}$ has been approximated by the energy difference $\Delta E_{red}^{gas} = -45.1$ kcal/mol obtained by the real-space grids method with UBLYP functional. Thus, it has been demonstrated that our novel approach, in which the excess electron has been regarded as a solute in the implementation of the QM/MM-ER method, can afford the reduction free energy in excellent agreement with the value obtained by the conventional approach. We conclude that the present method is adequate for the computation of the reduction free energy of coenzyme embedded in a protein.

REFERENCES

1. H. Takahashi, N. Matubayasi, M. Nakahara, and T. Nitta, *J. Chem. Phys.* **121**, 3989-3999 (2004).
2. N. Matubayasi and M. Nakahara, *J. Chem. Phys.* **113**, 6070-6081 (2004); *ibid.*, **117**, 3605-3616 (2002).
3. J. R. Chelikowsky, N. Troullier, and Y. Saad, *Phys. Rev. Lett.* **72**, 1240-1243 (1994).

A Theoretical Approach to a Novel Pathway of Proton Translocation of Cytochrome c Oxidase

Yu Takano and Haruki Nakamura

Institute for Protein Research, Osaka University, Suita, Osaka 565-0871, Japan

Abstract. A novel proton translocation pathway of cytochrome *c* oxidase (H pathway) has been suggested from high-resolution crystal structures. In the H pathway, the most interesting features are the proton transfer via a peptide bond and the control of proton translocation by the redox reaction of heme *a*. We have studied the two features of the H pathway with density functional calculations. The reduction of the proton affinity of heme *a* propionate is required to pass a proton to aspartate through a peptide bond, forming an enol peptide bond. The enol tautomer could be converted to the keto state by a proton wiring mechanism. The proton wiring keto–enol tautomerism is responsible for the unidirectional character of the proton translocation. The redox reaction of Fe ion of heme *a* changes the charge density on the formyl and propionate groups via the π conjugation of the porphyrin ring. Proton translocation would be coupled to the redox-induced change in the proton affinity of the heme *a* propionate.

Keywords: cytochrome *c* oxidase, density functional theory, proton wiring keto–enol tautomerism, proton translocation
PACS: 82.39Jn, 82.93Rt

INTRODUCTION

Cytochrome *c* oxidase (CcO), a terminal oxidase in the cell respiratory chain of mitochondria and aerobic bacteria, utilizes free energy of O_2 reduction to water for proton translocation across the inner mitochondrial membrane [1]. In spite of many studies of CcO [1–5], the molecular mechanism of the proton pumping is still unclear. Recently, crystal structures of bovine heart CcO in the fully oxidized and reduced states have determined at high resolution, suggesting a novel pathway for proton translocation (H pathway) [3, 4]. The H pathway shows two interesting features. One is the proton transfer via a peptide bond. The H pathway is composed of a hydrogen bond network and a water channel. The hydrogen bond network contains the peptide bond between Tyr440 and Ser441, which would facilitate unidirectional proton transfer (Figure 1). The proposed mechanism [3, 4, 6, 7] of the proton transfer via a peptide bond consists of the formation of an enol peptide bond followed by the keto–enol tautomerism of the peptide bond. The other is the control of proton translocation by the redox reaction of heme *a*.

In this study, we have investigated the two features of the H pathway of CcO using density functional calculations. First, a possible mechanism has been explored for the unique proton transfer [6]. Second, the electronic structures of heme *a* in the redox reaction has been examined.

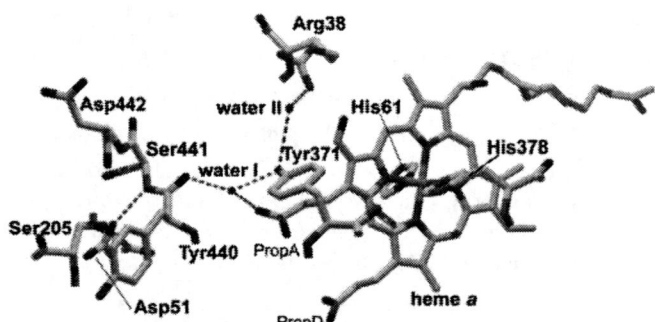

FIGURE 1. Hydrogen bond network in the proton transfer in the H pathway of CcO. The dotted lines show hydrogen bonding.

CP1046, *Selected Papers from ICNAAM 2007 and ICCMSE 2007*
edited by T. E. Simos, G. Maroulis, G. Psihoyios, and Ch. Tsitouras
© 2008 American Institute of Physics 978-0-7354-0574-5/08/$23.00

COMPUTATIONAL DETAILS

The reaction pathways for a proton transfer through the peptide bond were explored with B3LYP/6-31++G(d,p)//HF/3-21G calculations. Zero point energies and thermal corrections at 298 K (scaled by 0.9409) were included in the reported energies.

The proton transfer via a peptide bond was divided into two steps: the proton transfer forming an enol peptide bond and the keto–enol tautomerism of the peptide bond. Construction of the system was based on the crystal structure of fully oxidized CcO (PDB ID: 1v54) [3]. In the proton transfer forming an enol peptide bond, water I, the peptide bond, and Asp51 are concerned as proton donor, mediator, and acceptor, respectively (Figure 1). Tyr371 and the propionate A group (PropA) of heme a also take part in the proton transfer as hydrogen bond acceptors. As shown in Figure 2a, protonated water I was represented as a hydronium ion, the peptide bond was modeled as N-methyl acetamide, and acetate was used as the model of Asp51. Tyr371 was replaced with methanol because it works only as a hydrogen bond acceptor. Both acetate and acetic acid (Oγ1 is protonated) were utilized as models of the deprotonated and protonated PropA, respectively, because the protonation state of the PropA is still unknown. Two mechanisms are possible for the keto–enol tautomerism: a direct mechanism and an indirect mechanism. However, a direct mechanism shows high activation barrier due to the distorted transition structure. Since water molecules and amino acids surrounding the active site can facilitate proton transfer reactions, water molecules that could act as proton mediators were added to the tripeptide model (Tyr440–Ser441–Asp442) as shown in Figure 3a. To emulate the partially constraining effect of protein environment, we fixed the positions of the heavy atoms not directly involved in the reaction. These atoms are marked with asterisks in Figures 2a and 3a.

We employed UB3LYP exchange-correlation functionals to investigate the electronic structures of heme a. Basis sets used in all calculations were Tatewaki–Huzinaga MIDI (533(21)/53(21)/(41)) [7] plus Hay's d diffuse function ($\alpha = 0.1491$) for Fe ions [8], Pople's 6-31G(d) for C, O, and N atoms, and 6-31G for H atoms.

The geometries of heme a and axial ligands, His61 and His378, were taken from the three-dimensional atomic structure of fully oxidized and reduced CcO (Protein Data Bank, entry code 1v54 and 1v55, respectively) [3]. Methyl and vinyl group of heme a was substituted for hydrogen, and hydroxyfarnesylethyl group was replaced by methyl group. Imidazole was used as a model of His. H atoms were optimized with B3LYP method.

The environmental effect inside the protein was computed using the polarizable continuum model (PCM) with a dielectric constant of 4.0 at the HF/6-31++G(d,p) level.

All calculations were performed using GAUSSIAN03 program packages [8].

RESULTS AND DISCUSSION

Proton Transfer Forming an Enol Peptide Bond

The proton transfer forming an enol peptide bond was investigated at B3LYP/6-31++G(d,p)//HF/3-21G level in protein environment. The computed reaction pathways and the optimized structures are shown in Figures 2b and 2c. In the heme a model with the deprotonated PropA, geometry optimization of the reactant causes the spontaneous proton transfer to the Oγ2 atom of the PropA due to the strong electrostatic interaction between the protonated water I and the deprotonated PropA (Figure 2b), indicating that the PropA inhibits proton translocation. Even when amide proton of the mediating peptide bond is forced to be transferred to Asp51, this reaction is an endergonic reaction by 8.2 kcal/mol with the activation barrier of 11.5 kcal/mol, indicating that the deprotonation of PropA prevents the proton transfer. On the other hand, the energy profile is completely different in the model with the protonated PropA of heme a as illustrated in Figure 2c. In the optimization of the reactant, the proton moved to the peptide carbonyl oxygen, and an imidic acid [–C(OH)=N$^+$H–] is formed. The amide proton of the imidic acid is transferred to the carboxylate group of Asp51 with the activation barrier of 1.4 kcal/mol, forming an enol peptide bond [–C(OH)=N–]. The energy difference is exergonic by 2.6 kcal/mol, showing that the reverse transfer would easily occur.

A proton can exergonically transfer from water I to Asp51 via the peptide bond when the PropA of heme a is protonated. Protonation of PropA largely reduces the strength of the electrostatic interaction between the transported proton and the PropA. The deprotonated PropA, which is negatively charged, shows the strong electrostatic interaction with the hydronium ion and trapping the proton transfer to Asp51. The electrostatic interaction between the neutral protonated PropA and hydronium ion is weaker than that between Asp51 and hydronium ion. This energetic balance enables the proton to be transferred via the peptide bond. The protonation state of the heme a PropA influences the proton affinity of the Oγ2 atom in PropA.

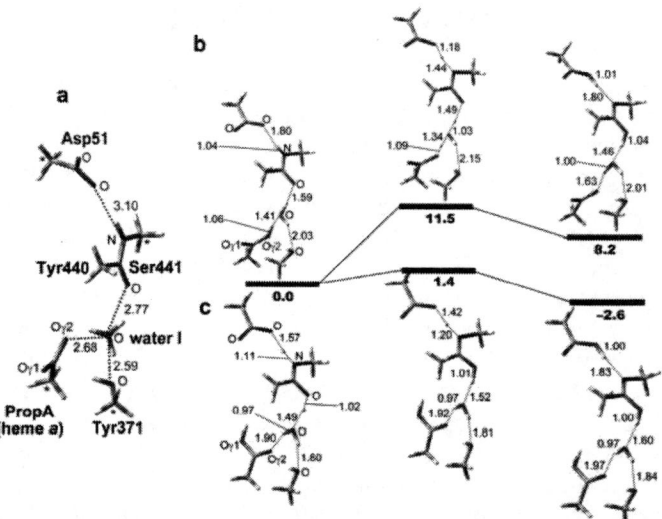

FIGURE 2. Model system (a), reaction profiles, and optimized stationary structures for the proton transfer forming an enol peptide bond for the deprotonated model (b) and the protonated model (c). The relative Gibbs free energies are shown in kcal/mol. Bond length are shown in angstroms.

Keto–enol Tautomerism of The Peptide Bond

We also explored the keto–enol tautomerism of the peptide bond in protein environment with the B3LYP/6-31++G(d,p)//HF/3-21G methods. Figure 3b shows the computed energy profile of the tautomerism and the optimized structures in the indirect keto–enol tautomerism. Three water molecules added as proton mediators and a neighboring peptide (Ser441–Asp442) catalyze the keto–enol tautomerism with the activation barrier to 21.8 kcal/mol, which is much smaller than that of the direct keto–enol tautomerism. The indirect keto-enol tautomerism requires energy more than the proton transfer forming an enol peptide bond does, indicating that keto-enol tautomerism is the rate-determining step. The activation barrier of the reverse reaction (keto to enol transition) is about 40 kcal/mol, blocking the keto to enol transition of the peptide bond. As a result, the peptide bond shows the unidirectional character of proton translocation.

In this computation, we employed water molecules as proton mediators in order to rationalize a mechanism of the indirect keto–enol tautomerism. However, what proton mediates the indirect tautomerism in the H pathway? We superimposed the computed transition structure on the X-ray crystallographic structure of the oxidized C*c*O and found Ser205 and protonated Asp51 near the mediating water molecules in the transition structure. No steric hindrance occurs between the X-ray structure and a water molecule linking the carbonyl oxygen of the enol peptide bond and amide nitrogen of the neighboring peptide bond. Ser205 and Asp51 can easily move to those positions because they are in the loop region of C*c*O. Since the pK_a of aspartic acid is lower than that of water, and Ser has comparable pK_a to water, we can expect that the activation barrier could decrease in the keto–enol tautomerism in the H pathway. In the H pathway, a water molecule, a neighboring amide group (Ser441–Asp442), Ser205, and protonated Asp51 could serve as proton mediators for the indirect proton wiring keto-enol tautomerism.

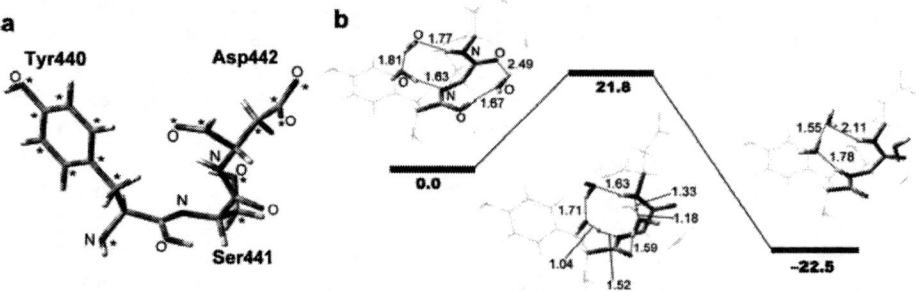

FIGURE 3. Model system, reaction profiles, and optimized stationary structures for keto-enol tautomerism of the peptide bond. The relative Gibbs free energies are shown in kcal/mol. Bond length are shown in angstroms.

Electronic Structures of Heme *a* in the Redox States

The electronic structures of heme *a* of C*c*O in the redox state have been investigated using the B3LYP method with polarizable continuum model. Figure 4 shows the difference charge density distribution of heme *a* in the redox reaction. The charge density on the d_{zy} orbital of Fe ion of heme *a* increases in the change from ferrous ion (Fe(II)) to ferric ion (Fe(III)), indicating that unpaired electron is occupied in the d_{zy} orbital. The redox reaction causes the charge density on the formyl group and PropA, which are involved in the hydrogen bond network, via the π conjugation of the porphyrin ring. It indicates that the redox reaction would be concerned with proton transfer in the hydrogen bond network.

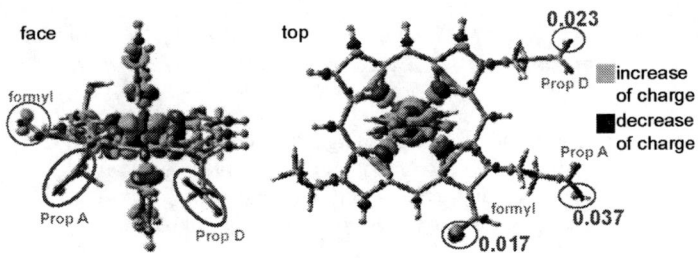

FIGURE 4. Difference charge density distribution of heme *a* in the redox reaction. Isosurface value is 0.002.

SUMMARY

In this paper, we have studied the two interesting features of the H pathway of C*c*O using density functional calculations. The proton transfer forming an enol peptide bond may require the protonation of the PropA of heme *a* to decrease the proton affinity of the Oγ2 atom. In this reaction, a proton moved to the carbonyl oxygen of the peptide bond, forming imidic acid followed by formation of an enol form of the peptide. The enol peptide bond would be converted to the keto form by a proton wiring mechanism using a water molecule, a neighboring amide group (Ser441–Asp442), Ser205, and protonated Asp51. Since the redox reaction of heme *a* also induces a change in the proton affinity of the PropA, electron transfer could be coupled to proton translocation through the redox-induced change in the proton affinity of the Oγ2 atom of the PropA in heme *a*.

ACKNOWLEDGMENTS

We are grateful to the Japan Society for the Promotion of Science for the Grant-in-Aid for Scientific Research (B) (18370063) and to the Ministry of Education, Culture, Sports, Science and Technology (MEXT) Japan for the Grant-in Aid for Scientific Research on Priority Areas "Structures of Biological Macromolecular Assemblies" (513-18054013) and Encouragement of Young Scientists (18750011). The computations were performed using Research Center for Computational Science, Okazaki, Japan and Cybermedia Center, Osaka University, Japan.

REFERENCES

1. S. Ferguson-Miller and G.T. Babcock, *Chem. Rev.* **96**, 2889-2907 (1996).
2. T. Tsukihara et al., *Science* **269**, 1069-1074 (1995).
3. T. Tsukihara et al., *Proc. Natl. Acad. Sci. USA* **100**, 15304-15309 (2003).
4. K. Shimokata et al., *Proc. Natl. Acad. Sci. USA* **104**, 4200-4205 (2007).
5. M. Antalik et al., *Biochemistry* **44**, 14881-14889 (2005).
6. Y. Takano and H. Nakamura, *Chem. Phys. Lett.* **430**, 149-155 (2006).
7. H. Tatewaki, S. Huzinaga, *J. Chem. Phys.* **72**, 399-405 (1980).
8. P. J. Hay, *J. Chem. Phys.* **66**, 4377-4384 (1977).
9. M. J. Frisch et al., GAUSSIAN03, Revision C.02, Gaussian Inc., Wallingford, CT, 2004.

Theoretical Studies on Stable Structure and Noncollinear Magnetism of Aluminum Clusters

Thi Viet Bac Phung, Kiyoshi Nishikawa, Hidemi Nagao

Division of Mathematical and Physical Science, Graduate School of Natural Science and Technology, Kanazawa University, Kanazawa 920-1192, Japan

Abstract. Electronic properties such as binding energy, magnetization of small Al_n (n=2-14) clusters are investigated using periodic density functional theory (DFT) calculations based on plane-wave. We present stable structures of Al_n clusters obtained from optimization with binding energy and magnetization. From noncollinear calculations of Al_3 and Al_5 clusters, we show the spin structure becomes delocalized. We discuss magnetic properties of Al_n clusters in relation to noncollinear calculations.

Keywords: clusters; spin polarization; density of states; collinear; noncollinear
PACS: 32.10.Dk, 73.22.-f

INTRODUCTION

Atomic clusters containing no more than a few hundred particles have been shown to display pronounced size-dependent properties such as geometric, electronic structure and "magic" number of clusters [1,2]. Investigations on structures, electronic properties are important due to their roles as building blocks for new nanomaterials, between molecular and bulk. Furthermore, some metallic clusters play important roles in many chemical reactions [3,4] as catalysis since of the structure of small metal clusters; therefore, there is a significant technological and fundamental motivation to understand the electronic and other properties of metallic clusters.

In studies of Al_n clusters, there are many experimental and theoretical results showing that the problems of stability and reactivity of these clusters show very complex [5-8]. The existing literature on Aluminum clusters has mainly focused on the structural and electronic properties. However, in spite of the experimental and theoretical effort, a consensus has not yet been achieved about the structure of many of the experimentally observed magic numbers [9].

In this work, by theoretical studies, we investigate stable structures properties of Al_n (n=2-14) clusters to understand their structural and electronic properties and discuss magnetic properties of their Al_n clusters in relation to noncollinear magnetism.

COMPUTATIONAL DETAILS

In this study, we have used the periodic program "Vienna ab initio simulation Package code" (VASP 4.6.31) [10] implemented the spin-polarized gradient-corrected DFT method [11]. The calculations have been done using the local density approximation (LDA) [12], generalized gradient approximation (GGA) [13] with the Perdew–Wang (PW91) [14] exchange-correlation functional for collinear magnetic moments. We have chosen the small core ultrasoft pseudopotentials (US-PP) and the plane wave basis sets supplied with VASP. Hence, the three 3s, 3p electrons for each Al are explicitly considered. For noncollinear calculation [15-18], the electron-ion interactions were described by the projector-augmented wave (PAW) method which proposed by Blochl and adapted by Kresse and Joubert [19-21]. The expansion includes all plane waves whose kinetic energies are less than energy cutoff E_{cut} = 161.5 eV and E_{cut} = 300.5 eV in US-PP and PAW, which ensure the convergence with respect to the basis set.

The periodic clusters are simulated by the repeated supercells in three directions. The periodic boundary conditions were imposed on a cubic cell with size length is 20Å (volume is 20Å×20Å×20Å), which is large enough

CP1046, *Selected Papers from ICNAAM 2007 and ICCMSE 2007*
edited by T. E. Simos, G. Maroulis, G. Psihoyios, and Ch. Tsitouras

for interaction between the clusters in neighboring cells to be negligible. The only Γ point is needed for this system described in this work, and spin-polarized calculations were performed for all calculations. In each calculation for Al_n clusters with n ≤ 6, the structure was fully relaxed until the difference of energy between two ionic steps relaxation less than 10^{-8}, for other clusters with the numbers of Al n ≥ 7, the convergence 10^{-6} has been used. All of these optimized structures were obtained by minimization of the total energy without symmetry constraints and using both Davidson algorithm for a few nonselfconsistent iterations and then to the RMM-DIIS algorithm as implemented in VASP (the generalized Kohn-Sham equations are solved employing a residual minimization scheme) to calculate the electronic ground state.

RESULTS AND DISCUSSIONS

Geometry and stability

The most stable optimized geometries of Al_n (n=2-14) clusters, bonding properties and magnetization are summarized in table 1 for comparison purposes. The binding energy per atom (E_B) is calculated by

$$E_B = \frac{n \times E_{Atom} - E_{Cluster}}{n}. \tag{1}$$

where E_{Atom} and $E_{Cluster}$ represent the total energy (eV) of a completely isolated Al atom and of the cluster, respectively. We use n to represent the number of atoms in the cluster. The unit of E_B is eV/atom.

The results of Aluminum dimer Al_2 with the bond distances in the range of 2.455Å -2.488Å (LDA/GGA in US-PP and GGA-PAW) and DOS shown in fig. 1 (a) indicates the electronic ground state is $(3p\sigma_u)^2$ $(^3\Sigma_g^-)$ agree well with available experimental, 2.466Å [5,22] and other theoretical data [6-8]. The ground state of and Al_3 in triangular arrangement with magnetization M = 1 as shown in fig. 1(b) was determined which is more stable than Al_3 linear structure, and the average bond length is 2.51Å.

From n ≥ 4, we found several optimized structures with planar and 3-dimentional clusters with different magnetizations as well. The planar structures of Al_4, Al_5 are slightly more stable than the 3-dimensional ones with the average bond lengths are 2.51 Å, 2.56 Å respectively. In contrast, the 3-dimentional octahedral structure of Al_6 cluster (M=2) was found to be much more stable than planar Al_6 (M=0), and this most stable structure of Al_6 is completely agreeable to the results by Pett et al [6]. The most stable structure in Al_7 cluster has the configuration as the combination of Al_3 and Al_4 in spatial structure.

TABLE 1. Binding Energy E_B (eV) and Magnetization M of the most stable Al_n (n=2-14) clusters.

Notation	Al_2	Al_3	Al_4	Al_5	Al_6	Al_7	Al_8
Structure							
E_B (eV/atom)	0.9370	1.4806	1.7107	1.9250	2.1358	2.3484	2.3728
Magnetization (M)	2	1	2	1	2	1	0

Notation	Al_9	Al_{10}	Al_{11}	Al_{12}	Al_{13}	Al_{14}
Structure						
E_B (eV/atom)	2.4239	2.4499	2.5017	2.5583	2.6883	2.7192
Magnetization M	1	2	1	2	1	0

(The values in the table 1 from US-PP/GGA calculation)

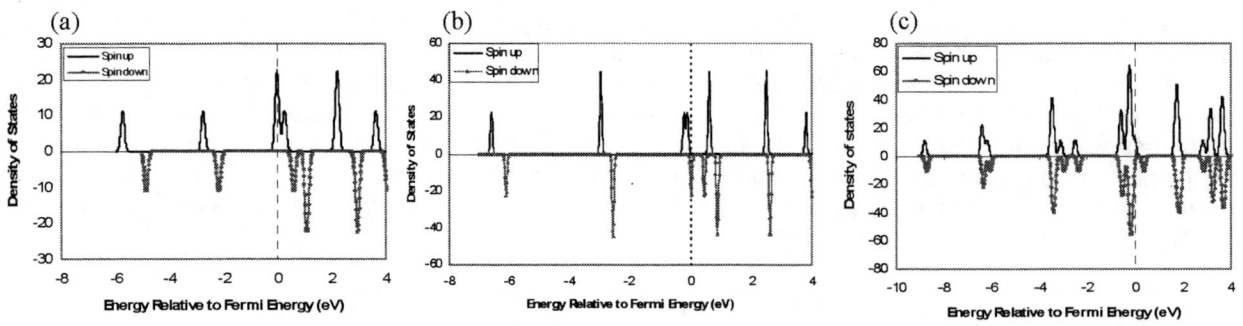

FIGURE 1. The spin-resolved total density of states (DOSs) of (a) Al_2, (b) Al_3 and (c) Al_{13} clusters, respectively. The dashed line represents the Fermi level.

In Al_8, there are so many similar minimized energies with the difference of magnetization were found and the most stable Al_8 structure was found with magnetization M=0 and binding energy 2.3728 (eV/atom) very closed to the value 2.3484 (eV/atom) of Al_7. The results of Al_n (n=9-13) stable structures and binding energies clearly indicated that the growth of clusters, the stable Al_{13} cluster has icosahedral structure close to an ideal icosahedron. This structure is similar to other metallic clusters were found recently [23,24], and the density of state of Al_{13} was plotted in fig. 1(c), M=1. And finally, Al_{14} stable structure was found as Al_{13} cluster and additional one Al atom and the interesting is this Al_{14} cluster likes Al_8 has magnetization equal to zero (M=0).

The results of magnetization suggest that magnetic moments are much dependent on particular clusters. Odd number of atoms has doublet spin structure, on the other hand, the ground state of even number shows triplet spin state. However, for the case of Al_8 and Al_{14} seems to be special structures with singlet spin state.

Noncollinear magnetism results

We performed firstly noncollinear magnetic properties of Al_3 and Al_5 clusters. The magnetic properties of these clusters are dominated by spin frustration. In table 2, the results for structure, binding energy, total magnetic moment of Al_3 and Al_5 clusters are presented, the arrows represent the magnetization orientation on each atom. (a) and (e) indicate the ground state in GGA-PAW collinear solution of Al_3 and Al_5 clusters with delocalized spin, respectively; (b) and (f) show noncollinear spin structure with frustrated possible spin configurations and binding energy per atom for their clusters. The orientation of the magnetic moment of (b) and (f) is determined to be minimum value of total magnetization of the system. Binding energy depends on the direction of spin on each atomic site. The binding

TABLE 2. Binding Energy E_B (eV/atom) and Total Magnetic Moment ($\mu_{B\,tot}$) of the Al_3 and Al_5 clusters.

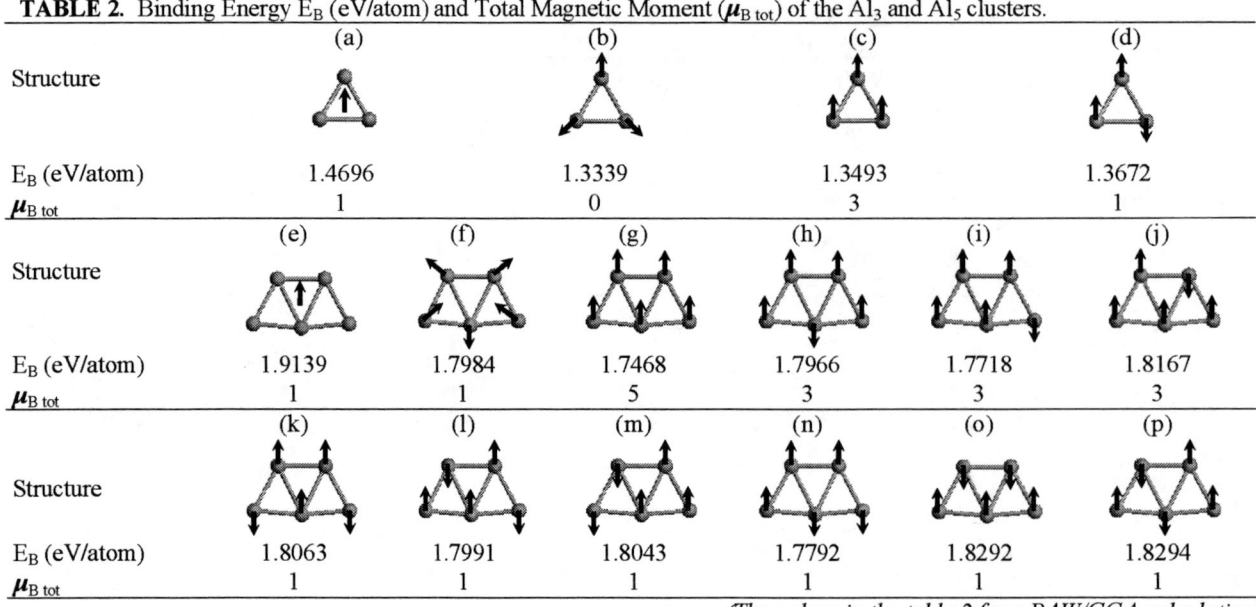

	(a)	(b)	(c)	(d)		
Structure						
E_B (eV/atom)	1.4696	1.3339	1.3493	1.3672		
$\mu_{B\,tot}$	1	0	3	1		
	(e)	(f)	(g)	(h)	(i)	(j)
Structure						
E_B (eV/atom)	1.9139	1.7984	1.7468	1.7966	1.7718	1.8167
$\mu_{B\,tot}$	1	1	5	3	3	3
	(k)	(l)	(m)	(n)	(o)	(p)
Structure						
E_B (eV/atom)	1.8063	1.7991	1.8043	1.7792	1.8292	1.8294
$\mu_{B\,tot}$	1	1	1	1	1	1

(The values in the table 2 from PAW/GGA calculation)

energy becomes lower than the ground state for each case. (d) of Al_3 and (k) - (p) of Al_5 clusters show the structures with possible localized spin configurations and binding energy with localized spin structure (μB tot = 1), and their biding energies of localized spin configuration become lower than collinear calculations. In Al3 cluster, the energy difference between the highest spin state (c) and the ground state becomes about 0.12 eV. On the other hand, that between (e) and (g) of Al_5 cluster becomes about 0.17 eV. In Al_5 cluster with $\mu_{B\ tot} = 3$, the spin structure of (h) is the most stable. These results suggest that the electronic state with localized spin structure becomes unstable and that distribution of spin structure in the ground state is more delocalized over the whole clusters.

CONCLUSIONS

The ground state structure of Aluminum clusters up to 14 atoms were determined using DFT with a plane wave basis set, not only collinear but also noncollinear magnetic arrangements were considered for small clusters. The dependence of magnetic properties on the particular systems in Aluminum clusters has been figured out, the magnetization of the even number Al_n clusters in ground state shows the triplet spin state and of the odd-sized ones has doublet spin, and the magnetization of Al_8 and Al_{14} vanishes. From noncollinear magnetic properties of Al_3 and Al_5 clusters, the distribution of spin structure in the ground state is delocalized.

Further, from these obtained electronic properties of small clusters, our results serve as a starting point for the prediction of the larger numbers of Al_n clusters, the critical cluster size at which the structural change occurs will be continuously studied and more comprehensive study of dependence of electronic properties on vibration mode in Aluminum clusters will be done in the following work.

ACKNOWLEDGMENTS

H.N is grateful for financial support of the Ministry of Education, Science and Culture of Japan (Research No. 19029014, No. 17064013)

REFERENCES

1. F. Baletto, R. Ferrando, *Rev. Mod. Phys.* **77**, 1 (2005).
2. J. Wu, F. Hagelberg, *J. Phys. Chem. A* **110**, 5901-5908 (2006).
3. M. Okumura, S. Tsubota, M. Haruta, *J. Mol. Cata. A. Chem.* **199**, 73-84 (2003).
4. M. Okumura, Y. Kitagawa, M. Hatura, K. Yamaguchi, *Applied Catalysis A* **291**, 27-44 (2005).
5. H.T. Upton, *J. Phys. Chem.* **90**, 754-759 (1986).
6. G. M. Pettersson, W. Bauschlicher, Jr. et al., *J. Chem. Phys.* **87**, 2205-2213 (1987).
7. A. Martinez, A. Vela et al., *J. Chem. Phys.* **101**, 10677-10685 (1994).
8. R. O. Jones, *Phys. Rev. Letters* **67**, 224-226 (1991).
9. E. G. Noya, J. P. K. Doye, F. Calvo, *Phys. Rev. B* **72**, 125407 (2006).
10. G. Kresse, J. Hafner, *Phys. Rev. B* **47**, 558 (1993) and VASP package webpage: *http://cms.mpi.univie.ac.at/vasp/*
11. P. Hohenberg and W. Kohn, *Phys. Rev.* **136**, B864 (1964).
12. W. Kohn and L. J. Sham, *Phys. Rev.* **140**, A1133 (1965).
13. J. P. Perdew, J. A. Chevary, S. H. Vosko, K. A. Jackson et al., *Phys. Rev. B* **46**, 6671 (1992).
14. J. P. Perdew, K. Burke, M. Ernzerhof, *Phys. Rev. Letters* **77**, 3865 (1996).
15. T. Oda, A. Pasquarello and R. Car, *Phys. Rev. Letters* **80**, 3622 - 3625 (1998).
16. D. Hobbs, G. Kresse, J. Hafner, *Phys. Rev. B* **62**, 11556-11570 (2000).
17. S. Yamanaka, D. Yamaki, Y. Shigeta, H. Nagao et al., *J. Quan. Chem.* **80**, 664 (2000).
18. J. E. Peralta, G. E. Scuseria, M. J. Frisch, *Phys. Rev. B* **75**, 125119 (2007).
19. G. Kresse, D. Joubert, *Phys. Rev. B* **59**, 1758 (1999).
20. G. Kresse, J. Furthmuller, *Phys. Rev. B* **54**, 11169 (1996).
21. P. E. Blochl, *Phys. Rev. B* **50**, 17953 (1994).
22. D. S. Ginter, M. L. Ginter, and K. K. Innes, *Astrophys. J.* **139**, 365 (1964).
23. M. Salazar-Villanueva et al., *J. Phys. Chem.* **110**, 10274-10278 (2006).
24. W. Zhang, H. Zhao and L. Wang, *J. Phys. Chem. B* **108**, 2140-2147 (2004).

Natural Orbital Analysis of Difference Density Matrix of Cyanide Fe(III) Porphyrins

Daisuke Yamaki[1,2], Masanori Suzuki[1] and Masahiko Hada [1,2]

[1] *Department of Chemistry, Graduate School of Science, Tokyo Metropolitan University, 1-1 Minami-Osawa, Hachioji, Tokyo 192-0397, Japan*
[2] *CREST, Japan Science and Technology Agency (JST), 4-1-8 Honcho Kawaguchi, Saitama 332-0012, Japan*

Abstract. In order to elucidate the difference of electronic states of bis-cyanide porphyrinato iron(III) and cyanide imidazole porphyrinato iron(III), the difference density matrix between the two molecules were calculated. The difference density matrix was analyzed by its natural orbitals. Contributions of Fermi contact term to the natural orbitals were evaluated. By replacement of the CN ligand in the bis-cyanide porphyrinato iron with the imidazole ligand, the negative spin density at ^{13}C in the cyanide was enhanced. It is found that the enhancement of the spin density in the porphyrin part arisen from the replacement was small. This is contrast to the spin densities themselves appeared in the porphyrin part as well as in the CN ligand.

Keywords: natural orbital analysis, difference density matrix, Fermi contact term of shielding constant, paramagnetic NMR, ab initio, cyanide porphyrinato iron.
PACS: 31.10.+z, 31.15.Ar

INTRODUCTION

From the first-order density matrix any one-electron properties can be obtained. Similarly, from the difference of first-order density matrices of two states, the difference of one-electron properties can be obtained. The difference density matrix is useful when the difference or change of states is focused on because the inert parts of the matrices are canceled out. By using the natural orbitals of difference density matrix, discussions based on occupation number and analysis of one-electron properties is possible. Previously, we dealt with differences of the second-order density matrices of two states of magnetic molecules and have shown the efficiency of the analysis for the results of ab initio calculations [1].

The porphyrinato iron complex has attracted much interest because they participate in biological reactions. For bis-cyanide porphyrinato iron(III) and cyanide imidazole porphyrinato iron(III) (the chemical structures shown in Figure 1), the ^{13}C NMR chemical shift of cyanide ($^{13}CN^-$) was observed[2]. By replacing the cyanide with the imidazole the paramagnetic NMR shift was significantly decreased. Accurate calculations of the electronic states and the paramagnetic NMR shift have been performed by Hada[3]. We focused on the difference of the electronic states of the two molecules. The Fermi contact term of the shielding constant is dominant for the paramagnetic NMR shift. The Fermi contact term is proportional to the spin density on the probe nucleus ^{13}C[4]. The paramagnetic NMR shift is strongly affected by the change of electronic states of the molecules. The change of shift depends on difference of the 1st order density matrices. The application of the difference density matrix analysis to these molecules is interesting.

In order to elucidate the difference of electronic states of bis-cyanide porphyrinato iron(III) and cyanide imidazole porphyrinato iron(III), we applied the difference density matrix analysis to the molecules, and calculated the contributions of Fermi contact term to the natural orbitals.

CP1046, *Selected Papers from ICNAAM 2007 and ICCMSE 2007*
edited by T. E. Simos, G. Maroulis, G. Psihoyios, and Ch. Tsitouras
© 2008 American Institute of Physics 978-0-7354-0574-5/08/$23.00

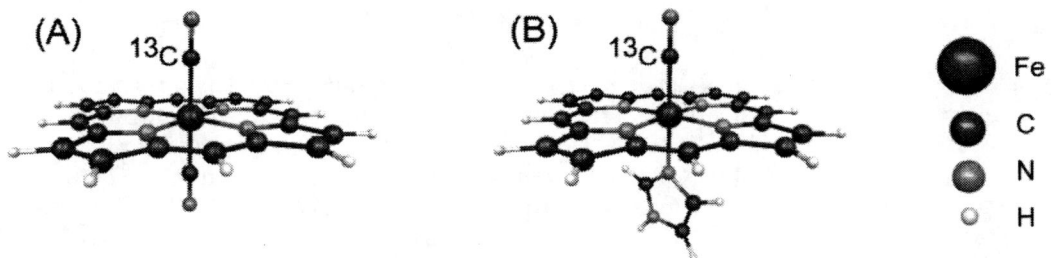

FIGURE 1. Chemical structures of (A) bis-cyanide porphyrinato iron(III) and (B) cyanide imidazole porphyrinato iron(III)

METHOD

The ordinary natural orbitals ϕ_i are obtained from the 1st order density matrix ρ by means of the diagonalization:

$$\rho(1',1) = \sum_i n_i \phi_i^*(1')\phi_i(1), \tag{1}$$

where n_i denotes occupation numbers. One-electron properties can be decomposed into the expectation values of the natural orbitals. The natural orbitals are the only basis that can decompose generally. The natural orbitals of difference density matrix φ_i has similar feature to the ordinary natural orbitals.

$$\Delta\rho(1',1) = \rho^{State1}(1',1) - \rho^{State0}(1',1) = \sum_i d_i \varphi_i^*(1')\varphi_i(1), \tag{2}$$

where d_i denotes occupation numbers that indicate difference of occupations between the two states or molecules. Differences of one-electron properties can be decomposed into the expectation values of the operator of the property to natural orbitals. The difference density matrix is useful when the difference or states are focused on because the inert parts of the matrices are canceled out.

The Fermi contact term of the NMR shielding constant depends on the spin density matrix $\rho_S = \rho_\alpha - \rho_\beta$ [4], therefore the spin density matrix was used instead of the ordinary density matrix:

$$\sigma_A^{FC} = 4\pi g^2 \beta^2 \langle \Psi | \hat{S}\delta_A | \Psi \rangle / k_B T \propto tr\left[\delta_A \rho_s\right], \tag{3}$$

where δ_A is the delta function at ^{13}C nucleus of the CN ligand. The natural orbital of the spin density matrix is referred to as the spin natural orbital [5]. These orbitals have been used as an efficient basis for the CI expansion to calculate the Fermi contact coupling in the literature [6]. The Fermi contact term can be decomposed into the natural orbitals of ρ_S because it is a one-electron property:

$$\sigma_A^{FC} \propto tr\left[\delta_A \rho_s\right] = \sum_i n_i \langle \phi_i | \delta_A | \phi_i \rangle = \sum_i n_i \langle \delta_A \rangle_i \tag{4}$$

where n_i denote spin occupation numbers. The summation of the products of the expectation values and the spin occupation numbers provides the total value of the Fermi contact term. The natural orbital with the large product is important for the property.

The difference of the property can be expanded by the natural orbitals of difference density matrix. For the Fermi contact term, the difference of spin density matrix was used:

$$\Delta\sigma_A^{FC} \propto tr\left[\delta_A \left\{\rho_s^{NC-Fe-Imi} - \rho_s^{NC-Fe-CN}\right\}\right] = \sum_i d_i \langle \varphi_i | \delta_A | \varphi_i \rangle = \sum_i d_i \langle \delta_A \rangle_i. \tag{5}$$

NUMERICAL DETAILS

The geometry was optimized for the cyanide imidazole porphyrinato iron. For the bis-cyanide porphyrinato iron, the same geometry was used except imidazole part. The other cyanide was placed symmetrically with respect to the iron atom. The B3LYP method with basis-set (6221/6211/311) + diffuse-sp + f-functions for Fe atom, (721/41) for CN ligands, (651/51) + d-functions for N atoms in the porphyrin and imidazole and (651/51) for C atoms in the porphyrin and imidazole was used to obtain density matrices of the electronic states. This basis set is same as the literature [3]. The electronic states calculations were performed by using Gaussian03 [7]. The visualizations of orbitals were done by the Macmolplt program package [8].

RESULT AND DISCUSSION

First the natural orbitals of the spin density matrix of the two molecules: bis-cyanide porphyrinato iron(III) and cyanide imidazole porphyrinato iron(III) are shown. The spin natural orbitals were obtained by using Eq (1). The contribution of an orbital to the ^{13}C Fermi contact coupling was obtained by the product of the spin occupation number and spin density at the ^{13}C nucleus: $n_i < \delta_A >_i$.

Figures (A1-3) and (B1-3) in Figure 2 show the spin natural orbitals of the two molecules. The Figures (A1) and (B1) shows spin natural orbitals with the largest occupation numbers. The spin occupation numbers were almost 1. These orbitals were localized in the Fe atom. This means that these orbitals are the spin sources of the molecules. Because the node of the orbitals included the probe nucleus ^{13}C, they have no contribution to the Fermi contact term.

Figures (A2) and (A3) show the orbitals of the bis-cyanide porphyrinato iron with the two largest contributions of the Fermi contact term. The contributions were -13.1 x 10^{-3} a.u. and -8.5 x 10^{-3} a.u. respectively. These orbitals were located in the proximal atoms from the Fe atom and had negative spin occupation numbers. These orbitals are responsible for negative spin densities induced by the spin polarization at C atom in the CN ligand and N atom in the porphyrin.

Figures (B2) and (B3) show the spin natural orbitals of the cyanide imidazole porphyrinato iron with the two largest contributions to the Fermi contact term. Different from (A2) and (A3), the spin polarization effects concerned with the Fermi contact term at the imidazole were smaller than the other proximal atoms from the Fe. The same as (A2) and (A3), the spin polarization effects were seen in the porphyrin part as well as the CN ligand.

Next the natural orbitals of different density matrix between the two molecules were calculated. The spin natural orbitals were obtained by using Eq (2). The contributions of the orbitals were calculated by $d_i < \delta_A >_i$.

Figures (C1) and (C2) show the orbitals with largest spin occupation numbers in the absolute value. These occupation numbers were localized at the Fe atoms and similar to the spin source orbitals. Similarly, these orbitals have no contribution to the Fermi contact coupling term because of the node.

Figure (C3) shows the orbital with the largest contributions, $d_i < \delta_A >_i$ = -23.3 x 10^{-3} a.u. to the Fermi contact term. This is the only dominant orbital for the change of the Fermi contact term because the next largest value of the contribution was $d_i < \delta_A >_i$ = -1.9 x 10^{-3} a.u., which is less than 1/10 of the value of the dominant orbital. The robe of this orbital at the CN ligand including ^{13}C was larger than the robes in the porphyrin. The enhancement of the spin density in the porphyrin arisen from the replacement was small. This is contrast to the spin densities themselves appeared in the porphyrin part as well as in the CN ligand.

CONCLUSION

The ab initio electronic structure calculations were performed for the bis-cyanide porphyrinato iron(III) and the cyanide imidazole porphyrinato iron(III). The difference density matrix was analyzed by its natural orbitals. Contributions of Fermi contact term to the natural orbitals were evaluated. By replacement of the CN ligand in the bis-cyanide porphyrinato iron with the imidazole ligand, the negative spin density at ^{13}C in the cyanide was enhanced. These results of the fermi-contact coupling term were consistent with the previous calculations [3]. From the natural orbital participated in this enhancement, it is found that The enhancement of the spin density in the porphyrin arisen from the replacement was small. This is contrast to the spin densities themselves appeared in the porphyrin part as well as in the CN ligand. The difference density matrix analysis is more efficient than the ordinary natural orbital analysis when the difference between two molecules is focused on.

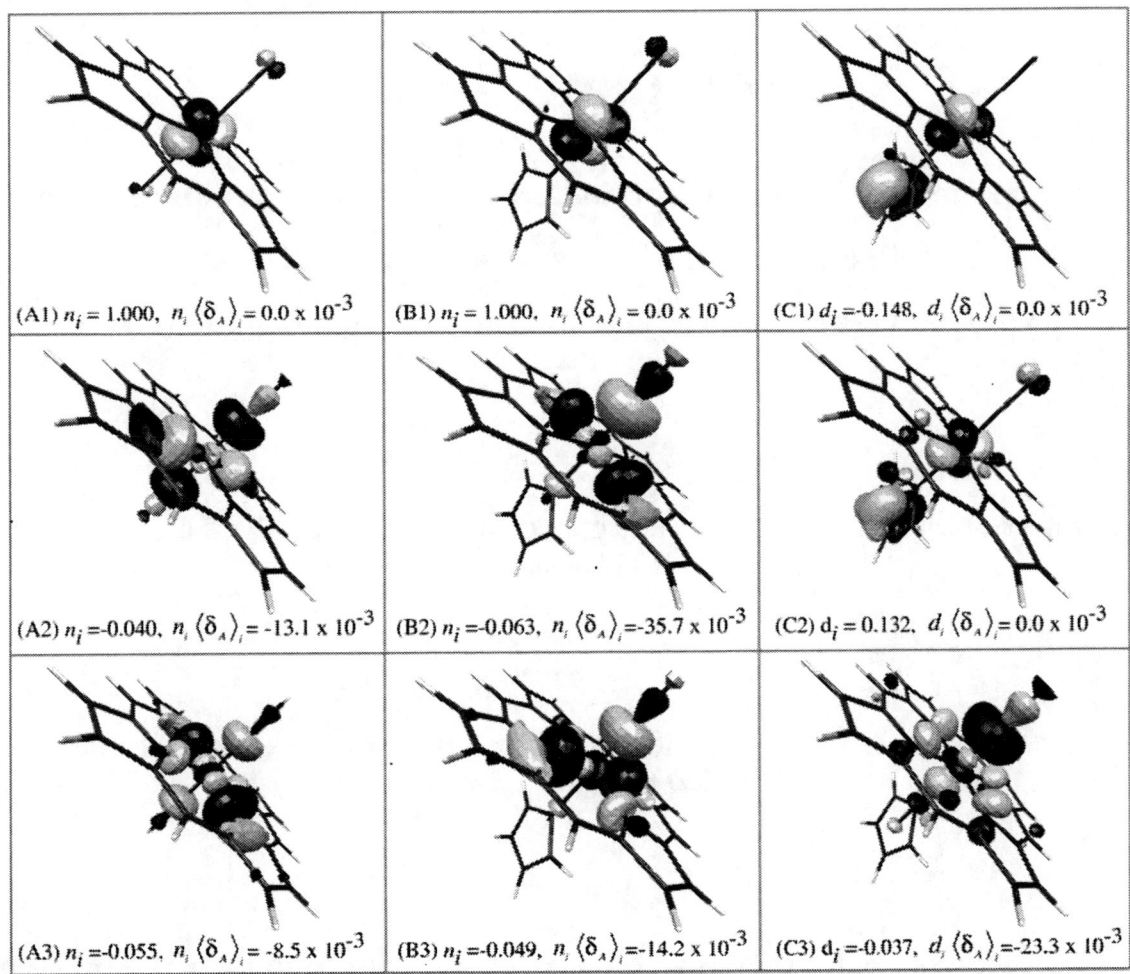

FIGURE 2. The spin natural orbitals of the density matrix of (A) the bis-cyanide porphyrinato iron and (B) the cyanide imidazole porphyrinato iron, and (C) the difference density matrix between the two. The only orbitals with largest spin occupation numbers or the largest contribution to the Fermi contact of ^{13}C in the CN ligand are shown.

REFERENCES

1. D. Yamaki, H. Nagao, K. Yamaguchi, Int. J. Quant. Chem., 85, 204 (2001)
2. H. Fujii, J. Am. Chem. Soc., 124, 5936 (2002).
3. M. Hada, J. Am. Chem. Soc., 126, 486 (2003)
4. Z. Rinkevicius, J. Vaara, L. Telyatnyk and O. Vahtras, J. Chem. Phys., 118, 2550 (2003).
5. D.M. Chipman, J. Chem. Phys. 78, 3112 (1983).
6. B. Engels, S.D. Peyerimhoff, Z. Phys. D - Atoms, Molecules and Clusters 13, 335 (1989).
7. Gaussian 03, Revision C.02, M. J. Frisch, G. W. Trucks, H. B. Schlegel, G. E. Scuseria, M. A. Robb, J. R. Cheeseman, J. A. Montgomery, Jr., T. Vreven, K. N. Kudin, J. C. Burant, J. M. Millam, S. S. Iyengar, J. Tomasi, V. Barone, B. Mennucci, M. Cossi, G. Scalmani, N. Rega, G. A. Petersson, H. Nakatsuji, M. Hada, M. Ehara, K. Toyota, R. Fukuda, J. Hasegawa, M. Ishida, T. Nakajima, Y. Honda, O. Kitao, H. Nakai, M. Klene, X. Li, J. E. Knox, H. P. Hratchian, J. B. Cross, V. Bakken, C. Adamo, J. Jaramillo, R. Gomperts, R. E. Stratmann, O. Yazyev, A. J. Austin, R. Cammi, C. Pomelli, J. W. Ochterski, P. Y. Ayala, K. Morokuma, G. A. Voth, P. Salvador, J. J. Dannenberg, V. G. Zakrzewski, S. Dapprich, A. D. Daniels, M. C. Strain, O. Farkas, D. K. Malick, A. D. Rabuck, K. Raghavachari, J. B. Foresman, J. V. Ortiz, Q. Cui, A. G. Baboul, S. Clifford, J. Cioslowski, B. B. Stefanov, G. Liu, A. Liashenko, P. Piskorz, I. Komaromi, R. L. Martin, D. J. Fox, T. Keith, M. A. Al-Laham, C. Y. Peng, A. Nanayakkara, M. Challacombe, P. M. W. Gill, B. Johnson, W. Chen, M. W. Wong, C. Gonzalez, and J. A. Pople, Gaussian, Inc., Wallingford CT, 2004.
8. Bode, B. M. and Gordon, M. S. J. Mol. Graphics Mod., 16, 133 (1998), http://www.scl.ameslab.gov/MacMolPlt/

CORRIGENDUM

"Elongation method applied to aperiodic systems - random polypeptides, high spin alignment, polymer in solvent, and DNA"

[AIP Conference Proceedings Volume 963, pp. 120-137 (2007)]

by

Y. Aoki, F. L. Gu, and Y. Orimoto

Department of Material Sciences, Faculty of Engineering Sciences, Kyushu University, 6-1 Kasuga-Park, Fukuoka 816-8580, Japan

S. Suhai

Molecular Biophysics, German Cancer Research Center (DKFZ), Im Neuenheimer Feld 580, D-69120 Heidelberg, Germany

A. Imamura

Hiroshima Kokusai Gakuin University, Nakano, Aki-ku 6-20-1, Hiroshima 739-0321, Japan

The affiliation of the last author (**A. Imamura**) in the above paper [see AIP Conference Proceedings Volume 963, COMPUTATIONAL METHODS IN SCIENCE AND ENGINEERING: Theory and Computation: Old Problems and New Challenges. Lectures Presented at the International Conference on Computational Methods in Science and Engineering 2007 (ICCMSE 2007): VOLUME 1, pp. 120-137.] has been erroneously typed and should be corrected to

"A. Imamura, Hiroshima Kokusai Gakuin University, Nakano, Aki-ku 6-20-1, Hiroshima 739-0321, Japan".

CP1046, *Selected Papers from ICNAAM 2007 and ICCMSE 2007*
edited by T. E. Simos, G. Maroulis, G. Psihoyios, and Ch. Tsitouras
© 2008 American Institute of Physics 978-0-7354-0574-5/08/$23.00

ICNAAM 2007

Algebras Generated by Geometric Scalar Forms and their Applications in Physics and Social Sciences

Jaime Keller

Departamento de Física y Química Teórica,
Facultad de Química Universidad Nacional Autónoma de México
AP 70-528, 04510, México D.F., MEXICO
(permanent address) and
Center for Computational Materials Science, General Physics
Technical University of Vienna, Gumpendorferstrasse 1A, A-1060, Vienna, Austria
keller@servidor.unam.mx , keller@cms.tuwien.ac.at

Abstract. The present paper analyzes the consequences of defining that the geometric scalar form is not necessarily quadratic, but in general K-atic, that is obtained from the Kth power of the linear form, requiring $\{e_i;\ i = 1,...,N;\ (e_i)^K = 1\}$ and $\vec{d} = \sum_i x_i e_i$. We consider the algebras which are thus generated, for positive integer K, a generalization of the geometric algebras we know under the names of Clifford or Grassmann algebras. We then obtain a set of geometric K-algebras. We also consider the generalization of special functions of geometry which corresponds to the K- order scalar forms (as trigonometric functions and other related geometric functions which are based on the use of quadratic forms). We present an overview of the use of quadratic forms in physics as in our general theory, we have called START. And, in order to give an introduction to the use of the more general K-algebras and to the possible application to sciences other than physics, the application to social sciences is considered.

For the applications to physics we show that quadratic spaces are a fundamental clue to understand the structure of theoretical physics (see, for example, Keller in ICNAAM 2005 and 2006).

Keywords: Algebraic structures, multilinear algebra, scalar N-forms, Geometric Algebras, Multivectors, Unified Field Theory, START
PACS: 02.10.Hh, 02.10.De, 02.10.Ud, 02.10.Xm, 02.40.-k, 02.40.Gh

GEOMETRIC SCALAR FORMS

The primitive notion of geometry is related, as the name indicates, to the measure of figures which can be found or represented on a two dimensional plane in our daily experience. The generalization of these concepts to 3 or more base space dimensions may require a profound mathematical analysis but is otherwise immediate and almost intuitive. This is due to the fact that we can visualize the extension from 2 to 3 dimensions and mainly because there are the simple, real numbers, relations among given scalar forms. For this the basic Pythagoras theorem, stating that for a rectangular triangle $a^2 + b^2 = c^2$, can be generalized to a general diagonal quadratic form $d^2 = \sum_i x_i^2$. This is such a strong constriction that many of our geometrically related quantities are based on the use of quadratic forms. Otherwise, in a flat manifold with translation invariant tangent space, the basic geometrical quantity is what we call vector $\vec{d}$, a generalized number which in geometry is a linear form, $\vec{d}$ can be represented on a basis e_i in N dimensions $\{e_i;\ i = 1,...,N;\ (e_i)^2 = 1\}$ as $\vec{d} = \sum_i x_i e_i$ such that the square, on both sides, corresponds to the general diagonal quadratic form.

As mentioned in the abstract we analyze the consequences of defining that the geometric scalar form is not necessarily quadratic, but in general K-atic, that is obtained from the (integer > 0) K^{th} power of the linear form, now requiring $\{e_i;\ i = 1,...,N;\ (e_i)^K = 1\}$ with $\vec{d} = \sum_i x_i e_i$. A set of algebras for positive integer K is generated, a generalization of the geometric algebras we know under the names of Clifford or Grassmann algebras: the geometric K-algebras. We shortly consider the geometries obtained when K is positive but not necessarily integer.

Natural Sciences are a useful description of nature within the scientific method. Social Sciences are, correspondingly, a useful description of social phenomena within the scientific method. Our basic consideration here is that the best description would be within a mathematical, geometrical, framework.

For the applications to physics we show that quadratic spaces are a fundamental clue to understand the structure of theoretical physics (see Keller in ICNAAM 2005 and 2006). The main theoretical structures of classical physics are

CP1046, *Selected Papers from ICNAAM 2007 and ICCMSE 2007*
edited by T. E. Simos, G. Maroulis, G. Psihoyios, and Ch. Tsitouras
© 2008 American Institute of Physics 978-0-7354-0574-5/08/$23.00

embedded within a 5-D flat quadratic space (3 dimensions of space **S**, one of time **T** and one of action **A**- like basis) manifolds (with Lorentz signature 1,4) faithfully providing a relativistic **R** theory **T** (START) describing the Newton and Maxwell formalisms, Geometrical Optics and General Relativity all of them as particular linear and quadratic forms of this (flat) START reference base space. The 5-D space has a quadratic form which enlarges the real quadratic form of the 4-D space–time $ds^2 = \overrightarrow{ds} \bullet \overrightarrow{ds}$ to a 5-D quadratic form $dS^2 = \overrightarrow{dS} \bullet \overrightarrow{dS}$. Physical phenomena correspond to either flat or curved 4-D surfaces embedded in this 5-D space. The study of the mathematical structure of Quantum Mechanics also requires the use of geometrical quadratic forms, the linear forms are called wave functions.

QUADRATIC AND K-ATIC SCALAR FORMS

Geometry originated from the study of 2-dimensional "geometric figures" as the triangle, the rectangle or the circumference where the idea of a N-dimensional metric became useful because scalar quadratic forms related the "length" of linear forms in orthogonal directions.

Quadratic Scalar Forms are intuitive "Natural" because the relation called **Pythagoras Theorem** is an **Experimental "Fact"**. The algebra that considers together the linear forms $\overrightarrow{d}$, their products and the scalar quadratic forms $d^2 = \overrightarrow{d} \cdot \overrightarrow{d}$, is the Clifford-Grassmann algebra, which is thus the algebra of geometry.

If the scalar form of the geometric algebra is $d^K = \overrightarrow{d} \cdot \overrightarrow{d} \cdot .... \cdot \overrightarrow{d}$ (integer K times) we obtain an extension we call K-algebra, which correspond to an extension of geometry to K-metry. Then $K - algebras \supset Clifford - Grassmann$ and $K - metry \supset geometry \; (K = 2)$.

Linear forms $\mathbf{x} = \sum_i x_i e_i$ with an N-elements basis set $\{e_i; i = 1, ..., N\}$ generate an algebra of multivectors $M = \sum_{i,j,k,...} x_i y_j z_k ... e_i e_j e_k ...$, provided some commutation relation is defined $e_i e_k = \exp(i2\pi/K) e_k e_i$ to give a meaning to the outer product of the basis vectors. If, moreover, an inner product of sets of K basis vectors is also introduced, $\langle e_i, e_j, ... \rangle_K \to \mathcal{K} \subset \mathbb{R}$, a (scalar) 0-form, a geometric algebra is obtained.

CLIFFORD ALGEBRA INTO K-ALGEBRA

Consider the quadratic (2,N) multivector algebra and analysis, where

Scalar **Quadratic** Form $\quad\Longleftrightarrow$	N Dimensional Base space
$F_{(2)} = d^2 = x_1^2 + x_2^2 + ... + x_N^2$	N Dimensional Linear Form

and the embedding algebra, (K,N)-algebra, where, for positive integer k

Scalar **K-atic Form** $F_{(K)} \quad\Leftrightarrow$	N Dimensional Base space
$\lvert d \rvert^K = \lvert x_1 \rvert^K + \lvert x_2 \rvert^K + ... + \lvert x_N \rvert^K$	N Dimensional Linear Form
(K-trigonometry, analysis...)	$(K = \frac{1}{k}, 1, k)$-metry (N Dim)

(The $K = 1$ can be considered as a quadratic metric through the definition of the square–root of a vector, a combination of spinors, see Keller 1991). The $1/k$ cases do not generate a faithful multivector algebra.

Many well known relations are Quadratic–Scalar–Form based, for example: $\frac{d\cos(\theta)}{d\theta} = -\sin(\theta)$ and $\frac{d\sin(\theta)}{d\theta} = \cos(\theta)$ would be generalized to $\frac{d\cos_K(\theta)}{d\theta} = -(\sin_K(\theta))^{K-1}$ and $\frac{d\sin_K(\theta)}{d\theta} = (\cos_K(\theta))^{K-1}$.

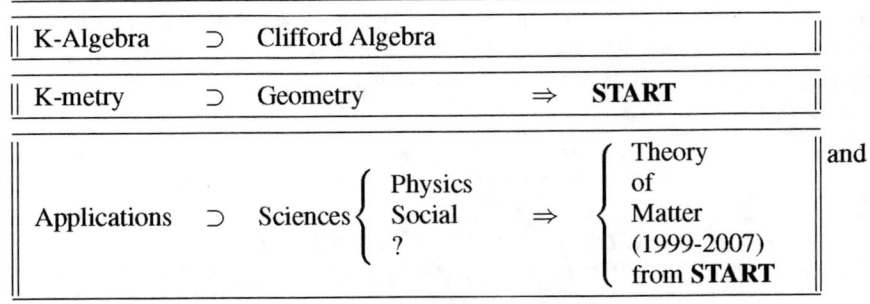

K-Algebra	$\supset$	Clifford Algebra		

K-metry	$\supset$	Geometry	$\Rightarrow$	**START**

Applications	$\supset$	Sciences $\begin{cases} \text{Physics} \\ \text{Social} \\ ? \end{cases}$	$\Rightarrow$	$\begin{cases} \text{Theory} \\ \text{of} \\ \text{Matter} \\ \text{(1999-2007)} \\ \text{from } \textbf{START} \end{cases}$ and

Geometry	$\Leftarrow$	$\mathbb{R}^N$ manifold	$\Rightarrow$	**K-metry**
Quadratic Form		Linear Form		K-order Form
(Scalar)	$\Leftarrow$	with	$\Rightarrow$	(Scalar)
N-dimensional		N Basis Vectors		N-dimensional
$e_i e_j = e^{\mathbf{i}\frac{2\pi}{2}} e_j e_i$		(base space)		$e_i e_j = e^{\mathbf{i}\frac{2\pi}{K}} e_j e_i$
$\Downarrow$		$\Updownarrow$		$\Downarrow$
Pythagoras Theo.		Multivectors		K-metric Theorem
		m-vectors	$\Rightarrow$	K-Algebra
		m=0,1,..,N$\times$(K-1)		
Trigonometry				K-Trigonometry
$\cos^2\theta + \sin^2\theta = 1$				$|\cos\theta|^K + |\sin\theta|^K = 1$
Clifford Algebra	$\Leftarrow$	(Grassmann, K=2)	$\Rightarrow$	[K=2$\Leftarrow$Physics]
Physics (**START**)				Social Sciences, etc

- The algebra has thus two basic numbers to define its dimension: the dimension N of the basis set and the dimension K of the number of elements to be multiplied together to obtain a scalar.

- If the dimension K refers to the order of the power of $[e_i]^K$ to obtain the scalar we will say that we have a K-atic algebra, the best known example is when the scalar form is a quadratic expression; these algebras are said to have a metric which in general is either diagonal or at least symmetric.

- Otherwise if the dimension K refers again to the number of different basis vectors to be multiplied together in $\langle e_i, e_j, ... \rangle_K \subset \mathbb{R}$ (with $j \neq i$ and in general all subindexes different) then we obtain a simplectic algebra where the best known case is also when $K = 2$ and the metric in this case is antisymmetric.

- New Functions are generated. In the present paper we define these sets of algebras, give the commutation relations for the algebras with a K-atic scalar form and relate the results to known and new examples.
  Motivation: Algebras and vector-multivector based functions allow new (geo)metric structures in physics and other sciences.

- K-order scalar forms, generate new relations for K-metric functions.

- K-algebra, allow the description of a larger collection of quantitative relations. New integrals and derivatives, new gauge groups.

- K-Trigonometry as a simple guiding example.

- Example of applications: START in physics (quadratic) and Theory of Government Action-Factual Benefits (K-atic). To begin consider **START** and quadratic scalar forms.

PRINCIPIA GEOMETRICA PHYSICAE (PHYSICS FROM START)

We study the physical objects as extended carriers of energy momentum and a set of properties we give the general name of charges. Our frame of reference is a **3+1+1 dimensional** quadratic manifold: **space-time and action**, all defined through their mathematical properties. The general structure is

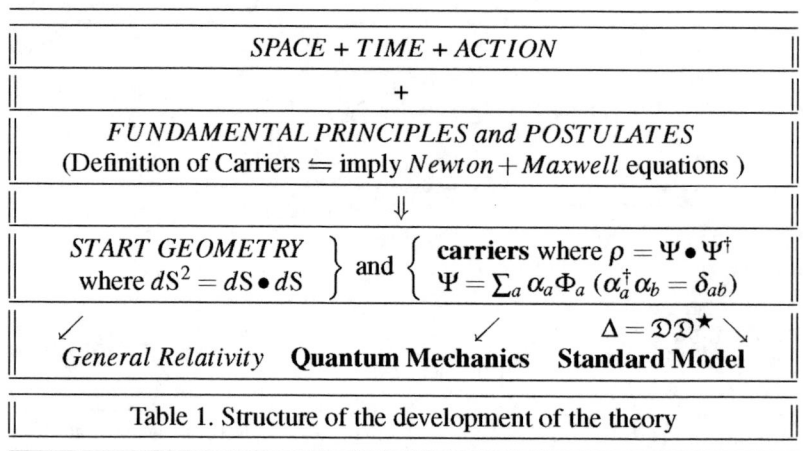

Table 1. Structure of the development of the theory

where $d\vec{S} = cdt\,e_0 + dl_x e_1 + dl_y e_2 + dl_z e_3 + K_0 da\,e_4$ is a space–time–action differential, Ψ a describing wave function obeying $\mathfrak{D}\Psi = 0$ and Δ a 5-D Laplacian.

- We have called START a theory where the purpose is to derive the fundamental laws of physics from a systematic use, as an application to physics, of the mathematical structure which can be constructed from a quadratic flat Lorentzian five dimensional manifold with signature $g_{UV} = diag(1, -1, -1, -1, -1)$ with $\{U, V\} = 0, 1, 2, 3, 4$, where the first 4 coordinates correspond to a flat (Minkowski) space–time, the fifth coordinate ($U = 4$) to action in reduced, equivalent distance, units.

- The group of symmetries related to the invariance of the quadratic form induces an augmented special relativity theory structure. The acronym of this base space is then START. From the definition of g_{UV} we derive the corresponding Clifford Algebra $\mathscr{C}_{1,4}$, and present its matrix representation.

- We study in this space the homogenous second order wave equations $KK^*\Phi = \mathfrak{K}^2\Phi = \partial^U \partial_U \Phi = 0$ and its set of linearizations $K\Psi = 0$ through $\Psi = K^*\Phi$. The properties and symmetries of this linear equation are shown to faithfully correspond to the structure of the Standard Model of elementary particles. New features are found.

- The acceptance of space and time as primitive concepts, parameterized by a 4-D manifold (x, y, z, t say, or x_μ; $\mu = 0, 1, 2, 3$), is now universal in physics. Otherwise material bodies (and radiation for that purpose) are always introduced as additional concepts.

- In our approach an extra "dimension" a is introduced to represent matter and energy (ME). This new variable will describe our primitive concept that ME is distributed in space and that this distribution evolves in time, tautologically distances and time are defined in relation to matter distribution and evolution.

- In our approach the tautology space-time-matter induces hypersurfaces $a = a(x, y, z, t)$. This is analogous to the definitions $x = x(t)$, $y = y(t)$ and $z = z(t)$ to study motion.

- The partial derivatives $\partial a/\partial x = p_x$, $\partial a/\partial y = p_y$, $\partial a/\partial z = p_z$ and $\partial a/\partial t = \varepsilon$ define a through its properties considering $\{\{\varepsilon, p_x, p_y, p_z\} \Longrightarrow \{p_\mu; \mu = 0, 1, 2, 3\}\}$ as energy–momentum.

- With this definitions matter–radiation distributions are represented mathematically as an equivalent action density

The 5-D quadratic manifold of Physics

The next step is to define a space–time–action quadratic 5-D manifold. Similar to the formal procedure to integrate space and time defining the quadratic Minkowski space, in our case we introduce a 5-D quadratic form. This is a generalization of the quadratic form which historically started with the Pythagorean formulation. Schematically:

Quadratic form of differential interval	Dim./Op.	Basic Group
$(dl)^2 = (dx)^2 + (dy)^2 + (dz)^2 \qquad (\triangle t)$	3-D ∇, ∇^2	Galileo
$ds^2 = (cdt)^2 - ((dx)^2 + (dy)^2 + (dz)^2)$	4-D $D, \square^2$	Poincaré
$dS^2 = (cdt)^2 - (dx^2 + dy^2 + dz^2) - dw^2$	5-D $\mathfrak{K}, \mathfrak{K}^2$	START

where we are defining for the action and for its differential $w = \kappa_{(0)} a$, $\kappa_{(0)} = d_{(0)}/h = c/E_{(0)}$, $dw^2 \equiv (dw)^2 = \sum_\mu \left(\kappa_{(0)} \left(\partial a / \partial x_\mu \right) dx_\mu \right)^2$; $\mu = 0, 1, 2, 3$ here l, (x, y, z), c, h, w, a, $E_{(0)}$ are **distance**, distance components, vacuum speed of light, Planck's constant, distance equivalent to action, action of a system and characteristic energy of the system respectively. Notice that in a physical description of motion where $x_\mu = x_\mu(t)$ we would consider $\left(dx_\mu \right)^2 = \left(\left(\partial x_\mu / \partial t \right) dt \right)^2 = \left(v_\mu dt \right)^2$ using the velocities v_μ as fundamental quantities and time as the basic parameter (above $\left(\partial a / \partial x_\mu \right) = p_\mu$ energy–momentum defining ACTION in relation to space–time, and considering space–time as a fundamental frame of reference).

There is a nesting from the general *space–time–action* manifold {with quadratic form $Q^{(5)}_{Form}$ which induces a Clifford Algebra $Cl_{1,4}$} first when fixing the Observed System and, second, when fixing the Observer for the description:

$\mathbb{R}^5 \overset{Q^{(5)}{}_{Form}}{\Longrightarrow} Cl_{1,4}$ $Q^{(5)}_{Form} \quad dS^2 = g_{UV} dx^U dx^V = 0$ $U, V = 0, 1, 2, 3, 4$	general *space-time-action* *where **any** observer* *describes **any** system* as bundle of null trajectories
$\mathbb{R}^4 \overset{Q^{(4)}{}_{Form}}{\Longrightarrow} Cl_{1,3}$ e_4 fixed (system is defined)	general *space-time* *to describe **given** system*
$\mathbb{R}^3 \overset{Q^{(3)}_{Form}}{\Longrightarrow} Cl_3$ e_0 (observer) and e_4 fixed	*space where **given** observer* *describes **given** system*

$\mathbf{e}_i = e_0 e_i = e_4 e_0 e_4 e_i$ in a 3-D$\subset$4-D$\subset$5-D series of embedding.

Once the basic geometry is defined, we are then working in a 5-D flat base space, Physics is introduced through: the **START Relativity Principle; All trajectories describing physical objects are null for all observers,** $dS^2 = 0$. **The vacuum Speed of Light is c for all observers, and the START Description Principle; Physical objects are described by bundles of null trajectories The environment of physical objects is described by interaction potentials and chemical potentials.**

The null trajectories condition reduces the five dimensions to effective four dimensions. A non-zero chemical potential could allow the physical object to exchange energy with the environment. The interaction potentials describe energy-momentum being shared.

- Our geometrical structure allows the development of a new, deductive scheme, formalism. For this purpose we need a minimal set of FUNDAMENTAL PRINCIPLES and POSTULATES, in such a form to obtain a comprehensive theory for Physics:
- **START Relativity** (the laws of physics are invariant under a 5-D group of isometries which includes 5-D Lorentz transformations);
- **Existence** (physical objects are represented by energy densities);
- **Least Action** (physically acceptable 'trajectories' correspond to null, optimal, possible, trajectories in START);
- **Quantized Exchange of Action** (we can define systems or subsystems as those among which a quanta of action can be exchanged) and,

- **Choice of Descriptions** (we should allow all useful physical models to be employed and properly based in START).

1. Twofold central purpose of START: first to have a **description of nature**, the way we perceive it with our senses and experiments, useful in accordance with the Scientific Method. Second START is aimed to be a valid **general mathematical theory** for all the fundamental physical objects and the frame of reference we use for their description.

2. By construction Planck's constant h is used in the definition of the **equivalent distance** corresponding to action.

3. The 5-D START geometry is presented as a **flat Lorentzian manifold**, time-like oriented in which all matter and light rays correspond to null lines. Massive objects are described as bundles of null trajectories in START. Observers to 4-D time-like trajectories.

4. When interactions are included the trajectories of all physical objects remain to be **null trajectories** which can also be represented as null geodesics in a curved 4-D subspace embedded in the START manifold. The curvature will obey a set of equations of which optics in a refractive medium or Einstein's equations of general relativity are particular cases.

5. All names in theoretical physics (energy, momentum, action, charges, etc.) are given their usual connotation and relationships. New, additional, relationships are found though.

6. The frame of reference is space($\mathbf{x}$)–time(ct). The action variable ($w = ad_S/h$) of each system S under consideration is expressed in units of an equivalent distance d_S/h, where h is Planck's constant and $d_S = hc/E_S^{(0)}$ the system's Compton wavelength. c and h are universal, whereas x, t and w are observer's relative.

SCALAR K-ATIC FORMS AND THEIR LINEAR FORMS. FUNDAMENTAL DEFINITIONS. $\{K = 1, ...\}$ GENERATED MULTIVECTOR ALGEBRAS.

We come now to the geometric K-algebras. A linear form with an N-elements basis set $\{e_i; i = 1, ..., N\}$ generates an algebra which is that of multivectors, provided some commutation relation is defined to give a meaning to the outer product of the basis vectors. If, moreover, an inner product of sets of K basis vectors is also introduced, for a mapping $\langle e_i, e_j, ... \rangle_K \rightarrow \mathscr{K} \subset \mathbb{R}$ producing a 0-form, a geometric algebra is obtained. The algebra has thus two basic numbers to define its dimension: the dimension N of the basis set and the dimension K of the number of elements to be multiplied together to obtain a scalar. If the dimension K refers to the order of the power of $[e_i]^K$ to obtain the scalar we will say that we have a K-atic algebra, the best known example is when the scalar form is a quadratic expression; these algebras are said to have a metric which in general is either diagonal or at least symmetric.

We obtain the set of algebras generated by the linearization of the scalar forms of the type $z^K = a^K + b^K + \cdots$, a K power version of the Pythagoras formula in N dimensions. We want then to solve the problem of finding the linear form

$$\mathbf{v} = \sum_{i=1}^{N} a^i \mathbf{e}_i \tag{1}$$

in an N-manifold with basis set $\mathbf{e}_i$ which linearizes the scalar K-atic form

$$|\mathbf{v}|^K \doteq \sum_{i=1}^{N} |a^i|^K \subset \mathbb{R} \tag{2}$$

a non diagonal K-metric and, if needed, mixed positive and negative $|a^i|^K$ terms are also allowed transformations of (2). We denote $\mathbf{v}$ as a 'vector' and the products $a^i \mathbf{e}_i b^j \mathbf{e}_j c^k \mathbf{e}_k ...$, $i \neq j \neq k \neq i ...$, as multivectors.

For this, in order not to have non scalar cross-terms, we define the K-algebra$-$commutation relation from the null value of the sums of the non-diagonal terms in $\mathbf{vv}...\mathbf{v}|_{K \text{ factors}}$, it is

$$\sum_{(l=0)}^{(l=K-1)} P_n^{(l)} \left(a^i a^j ... a^k \right)_{K \text{ indexes, not all indexes equal}} = 0. \tag{3}$$

In the $N-$dimensional vector algebra, from the outer product of the basis vectors $\{\mathbf{e}_i \; ; \; i = 1, ..., N\}$ a set of m-vectors can be defined as

$$\left(\mathbf{e}_i \mathbf{e}_j ... \mathbf{e}_k \right)_{m \text{ vectors, all indexes different}}$$

and, in particular, to fulfill the condition (3), the bi-vectors should obey

$$\mathbf{e}_i \mathbf{e}_j = \exp(\frac{1}{K} 2\pi i) \mathbf{e}_j \mathbf{e}_i \,. \tag{4}$$

Notice that the use of the commutation rules requires, to fulfill the condition (3), to strictly follow the order in each factor of the analysis of the permutation procedure because

$$\mathbf{e}_i \mathbf{e}_j = \exp(\frac{1}{K} 2\pi i) \mathbf{e}_j \mathbf{e}_i \text{ implies that } \mathbf{e}_j \mathbf{e}_i = \exp(\frac{-1}{K} 2\pi i) \mathbf{e}_i \mathbf{e}_j$$

and therefore $\mathbf{e}_j \mathbf{e}_i = \exp(\frac{K-1}{K} 2\pi i) \mathbf{e}_i \mathbf{e}_j$ in the K-atic-scalar-form algebra.

The dimension of the multivector algebra is K^N.

Notice that $K = 2$ is again a special case where $K - 1 = 1$. The (4) fulfills (3) because there are respectively $0, 1, ..., l, ..., (K-1)$ binary permutations in each term of the summation (3) and we have the identity

$$\sum_{l=0}^{K} \exp(\frac{l}{K} 2\pi i) = 0$$

(the two dimensional representation of the set of K complex numbers $\exp(\frac{l}{K} 2\pi i)$ results in a symmetric $K-$star of unit concurrent force-like vectors, in equilibrium, with "zero net force" condition). This includes, as particular case, the Grassmann-Clifford algebras for quadratic forms where $K = 2$ and there are only the product of reference and one binary permutation.

The obvious examples are those where $K = 2$ which are the well known Grassmann- Clifford Algebra. We do not need to discuss the algebras as they are well documented in the literature [1, 2, 3, 4]. In this case, $K = 2$, we have $\exp(\frac{1}{K} 2\pi i) = -1$ and also $\exp(\frac{-1}{K} 2\pi i) = -1$ and the name 'commutation' became synonymous of the relation

$$\mathbf{e}_i \mathbf{e}_j = \exp(\frac{1}{K} 2\pi i) \mathbf{e}_j \mathbf{e}_i \Longrightarrow \mathbf{e}_i \mathbf{e}_j = -\mathbf{e}_j \mathbf{e}_i \Longrightarrow \mathbf{e}_j \mathbf{e}_i = -\mathbf{e}_i \mathbf{e}_j \tag{5}$$

where neither the value of K nor that we are using a $K-$root of unity is explicitly shown.

For the case $K = 3$ we find in the literature [5, 6, 7, 8] examples, usually called ternary algebras, with representations where $N = 1, 2, 3$ only [9]. It is possible that the reason for this lays in the fact that 3×3 matrix representations are used which only allow those values of N. and their similarity transformations, are the only available possibilities.

A similar situation, $N = K$, occurs with $K = 4$, usually called quartic algebras. A third case which is commonly found in the literature is that of the so-called q-deformed algebras

$$\exp(\frac{-1}{K} \pi i) \mathbf{e}_i \mathbf{e}_j = \exp(\frac{1}{K} \pi i) \mathbf{e}_j \mathbf{e}_i$$

where in general K is considered not just an integer but a real number. If K is a real number the dimension of the algebra would also be a real number from the point of view of the scalar form, otherwise it can be any dimension N with respect to the basis set $\{\mathbf{e}_i; i = 1, 2, ..., N\}$.

The algebras here described can be nested. Either for sets of the N basis vectors of for sets of values of $K^{(i)}$, $\sum_i K^{(i)} = K$.

SOCIAL SCIENCES

The type of problems in Social Sciences we consider are those which can be analyzed with geometrical-like mathematical structures. This can be illustrated with our theory **Government Action Benefits** formalized with a mathematical structure which can be represented by 2-D or higher dimensional graphs.Considering for example the geometrical representation of costs or benefits in

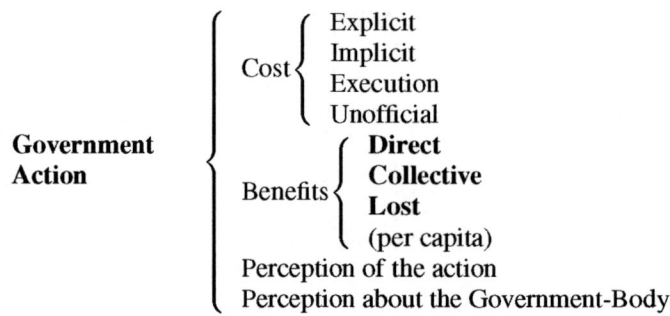

This study generated a theory of factual benefits[20]. Previous work referred to induced development of social bodies.

For the complicated cases that arise in these problems we found it convenient to nest K-algebras definitions for diferent values of K. This corresponds to create a series of K-algebras with K-atic forms with $K_1, K_2, \dots$ and then combining the resulting absolute values for the $\mathbf{E}$ vectors with another K-atic form where $K = K_0$. For example working in a x, y plane with some K-atic form $d^K = x^K + y^K$ and the resulting vector being combined with the composite z vector ($z^{K''} = p^{K''} + q^{K''}$) using a new K'-atic form $d^{K'} + z^{K'} = v^{K'}$.

ACKNOWLEDGMENTS

The technical support of TA Irma Aragón is gratefully acknowledged.

REFERENCES

1. W. E. Baylis, *Clifford (Geometric) Algebra with Applications to Physics, mathematics, and engineering*, Birkhäuser, Boston, 1996.
2. P. Lounesto, *Clifford Algebras and Spinor*, Cambridge University Press, Cambridge, 1996.
3. H. Grassmann, *Die Wissenschaft der extension Grösse oder die Ausdehnunglehre, eine neue mathematischen Disciplin*, Leipzig, 1844.
4. J. Keller, *Spinors and multivectors as a unified tool for spacetime geometry and for elementary particle physics. Int. Journal of Theoretical Physics*, **30**(2), 137–184 (1991) .
5. A. Cayley, *Cambridge Math. Journ.* **4**, 1 (1845).
6. J. J. Sylvester, *Johns Hopkins Circ. Journ.* **3**, 7 (1883).
7. M. Kapranov, I.M. Gelfand and A. Zelevinskii, *Discriminants, Resultants, and Multidimensional Determinants*, Birkhäuser, Boston, 1994.
8. L. Vainerman and R. Kerner, *Journal of Math. Physics* **37** (5), 2553 (1996).
9. A. Dubrovski and V. Guennadi, (LAPTH-06, CERN-HP-TH/2006-118, PNPI-06), *Ternary numbers and algebras, Reflexive numbers and Berger graphs*, *Advances in Applied Clifford Algebras*, to be published 2007.
10. J. Keller, *The Theory of the Electron A Theory of Matter from START*, From the series: Fundamental Theories of Physics, Kluwer Academic Publishers, Dordrecht, Vol. 115, 2001, e-bookhttp://www.wkap.nl/prod/b/0-7923-6819-3
11. J. Keller, *General Relativity from START*, in: *Clifford Analysis and Related Topics*, K. Gürlebeck X. Ji, W. Sprössig Eds., *Adv. in Applied Clifford Algebras* **11**(S2), 183-204 (2001).; Also J. Keller, "START: 4-D to 5-D generalization of the (Minkowski-)Lorentz Geometry", *Proceedings of the Meeting Lorentzian Geometry-Murcia-2003* in the series: Publicaciones de la Real Sociedad Matemática Española, **6** (2004).
12. J. Keller, General Relativity as a Symmetry of a Unified Space-Time-Action Geometrical Space, *Proceedings of Institute of Mathematics of NAS of Ukraine*, **43**(2) (2004); also J. Keller, A Derivation of Quantum Mechanics from START, *idem* **50**, edited by A.G. Nikitin, V.M. Boyko and R.O. Popovych (and I.A.Yehorchenko), Kyiv, Institute of Mathematics, (2002) pp. 557-568 and (2004) pp. 811-820; http://www.imath.kiev.ua/activities.php
13. J. Keller, Unification of Electrodynamics and Gravity from START, *Annales de la Fond. Louis de Broglie* **27**(3), 359 (2002). http://www.ensmp.fr/aflb/AFLB-272/aflb272p359.htm
14. J. Keller, A Comprehensive Theory of the Electron from START, in: *What is the Electron?*, in the series Modern Structures, Theories, Hypotheses, edited by V. Simulik, Montreal: Apeiron, 2005, pp.1-28.
15. J. Keller, "Principia Geometrica Physicae", Proceedings of the 6th International Symposium *Frontiers of Fundamental and Computational Physics-Udine (Italy), Sep. 26-29, 2004*, edited by Kluwer Academic Publishers, Boston, 2005 pp.163-173.

16. J. Keller, Unification of Electrodynamics and Gravity from START, *Annales de la Fondation Louis de Broglie* **27**(S), 359-410 (2002); J. Keller, *Advances in Applied Clifford Algebras* **11** (S2), 183-204 (2001); J. Keller, A Theory of the Neutrino from START, *Electromagnetic Phenomena* **3** (9),122-139 (2003).

17. J. Keller, Geometrical Principles of Physics, in: *International Conference on Numerical Analysis and Applied Mathematics-2005 (ESCMSE)*, edited by T. E, Simos, Wiley-VCH, 2005, pp.126-127; Ibid 2006.

18. J. Keller, "Inventio Principia Geometrica Physicae", in the Proceedings of the *7th International Conference on Clifford Algebras and their Applications -ICCA7, Toulouse, France, May 19-29, 2005*, edited by Pierre Anglés, Birkhauser, Basel, 2007.

19. J. Keller, Algebras Generated by Scalar K-atic Forms and Their Linear Forms, *Advances in Applied Clifford Algebras* (Online First Dec. 2006) **17** (2), 241-244 (2007).

20. J. Keller, Set of lectures on the general subject "Algebras Generated by Scalar K-atic Forms and Their Application to Study the Benefits of a Government Action", presented at *The Ixtapa 2007 Mathematical Methods Conference*, at *The Queretaro 2007 Hypercomplex Analysis Conference*, at *The Bedlewo 2007 Clifford Algebras and Physics Conference* and at *The Chelm 2007 Analytic Functions Conference*.

On Existence of Periodic Solutions for Certain Nonlinear Higher Order Autonomous Ordinary Differential Equations

M. Bayat, Z. Khatami and B. Mehri[1]

Department of Mathematics, Institute for Advanced
Studies in Basic Sciences 45195-1159, Zanjan, IRAN

Abstract. In this paper, we study the existence of periodic solutions for autonomous nonlinear ordinary differential equations of order n. Our method is based on the evaluation of Brouwer's degree theory and making use of the homotopy invariance property of the topological degree and also suitable norm inequalities. For this, we prove two lemmas about the second and third order ODE systems and then present two theorems about the sufficient conditions for the existence of periodic solutions for the even and odd n-order ODE respectively.

Keywords: Nonlinear equations, Periodic solutions, Topological degree, Brouwer's degree theory, Homotopy invariance.

INTRODUCTION

It is interesting to note that the existence of periodic solutions of nonlinear autonomous differential equations has not been extensively investigated. The Poincare-Bendixon theorem, which is a powerful tool for the investigation of periodic solutions of second order differential equations, is not applicable for third and higher order systems. In what follows, we use the idea of Brouwer's degree theorem [1,2] to prove the existence of periodic solutions of higher order systems.

The existence of nontrivial periodic solutions of autonomous nonlinear third, fourth and fifth order differential equations of the forms:

$$x''' + f(x, x', x'') = 0, \qquad f(x, -x', x'') = -f(x, x', x''),$$

$$x^{(4)} + f(x, x', x'', x''') = 0, \qquad f(-x, x', -x'', x''') = -f(x, x', x'', x'''),$$

$$x^{(5)} + f(x, x', x'', x''', x^{(4)}) = 0, \quad f(x, -x', x'', -x''', x^{(4)}) = -f(x, x', x'', x''', x^{(4)}).$$

has been investigated in [3-7].

MAIN RESULTS

For a closed domain D, define

$$|F(u)|_D = \max_{u \in D} |f(u)|,$$

$$\|F(u)\|_D = \max_{u \in D} |f(u)|_\infty.$$

Lemma 1. Let us consider the following second order system:

$$X'' + F(X, X') = 0, \qquad (X \in \mathbb{R}^n) \tag{1}$$

where n is an odd positive integer and F is smooth enough to ensure the uniqueness and existence of solutions. Consider $c > 0$ and diagonal matrix A as:

$$A = \text{diag}(a_1, \cdots, a_n), \qquad a_i > 0,$$

[1] bayat@iasbs.ac.ir, zkatami11@yahoo.com, mehri@iasbs.ac.ir

CP1046, *Selected Papers from ICNAAM 2007 and ICCMSE 2007*
edited by T. E. Simos, G. Maroulis, G. Psihoyios, and Ch. Tsitouras
© 2008 American Institute of Physics 978-0-7354-0574-5/08/$23.00

$$1 \le \frac{a_m}{a_M} < 2, \qquad a_m = \min_{1 \le i \le n} \{a_i\}, \qquad a_M = \max_{1 \le i \le n} \{a_i\},$$

and by considering $\|C\|_\infty = \max\{|c_i| : i = 1, \cdots, n\}$ for $C \in \mathbb{R}^n$, we define,

$$D = \left\{ (X, X') : \|X\|_\infty \le 2c, \|X'\|_\infty \le 2ca_M \right\},$$

$$\omega_0 = \frac{\pi}{a_m + a_M}, \qquad \omega_1 = 3\omega_0.$$

Now, if we have $\delta c > \frac{3M}{a_m^2}$, where

$$\delta = \min\{|\sin(a_i\omega_j)| : i = 1, \cdots, n, j = 0, 1\},$$

$$M = \max\{\|A^2 X - F(X, X')\|_\infty : (X, X') \in D\},$$

then there exists ω, with $\omega_0 < \omega < \omega_1$ such that for a proper initial conditions, the above system has a solution which satisfies the following boundary conditions:

$$X(0) = X(\omega) = 0.$$

Now, we generalize this result for third order systems as follows:

Lemma 2. Consider the third order system with the same assumptions on the existence and uniqueness of the solutions as Lemma 2.1:

$$X''' + F(X, X', X'') = 0, \qquad\qquad (X \in \mathbb{R}^n), \qquad\qquad (2)$$

with

$$D = \left\{ (X, X', X'') : \|X\|_\infty \le \frac{2c}{a_m}, \|X'\|_\infty \le 2c, \|X''\|_\infty \le 2ca_M \right\}.$$

If we have $\delta c > \frac{3M}{a_m^2}$, then there exist ω $(\omega_0 < \omega < \omega_1)$, such that

$$X'(0) = X'(\omega) = 0.$$

Now, we consider the following equations of order $2n$:

$$x^{(2n)} + f(x, ..., x^{(2n-1)}) = 0 \qquad\qquad (3)$$

with the assumption of existence and uniqueness of the solutions.

Theorem 1. If there exist $a_i, i = 1, ..., n$ and a closed domain D of the origin in $\mathbb{R}^{2n}$ such that

$$M = \left| \sum_{i=0}^{n-1} a_i x^{(2i)} - f(x, ..., x^{(2n-1)}) \right|_D$$

is small enough and furthermore the following parity condition holds:

$$f(-x, x', ..., -x^{(2i)}, ..., x^{(2n-1)}) = -f(x, x', ..., x^{(2n-1)}),$$

then system (2.1) has a periodic solution.

Now, We consider the equations of order $2n + 1$ as follows:

$$x^{(2n+1)} + f(x, ..., x^{(2n)}) = 0 \qquad\qquad (4)$$

with the assumption of existence and uniqueness of the solutions.

Theorem 2. If there exist $a_i, i = 1, ..., n$ and a closed domain D of the origin in $\mathbb{R}^{2n+1}$ such that

$$M = \left| \sum_{i=0}^{n-1} a_i x^{(2i+1)} - f(x, ..., x^{(2n)}) \right|_D$$

is small enough and furthermore the following parity condition holds:

$$f(-x, x', ..., -x^{(2i+1)}, ..., x^{(2n)}) = -f(x, x', ..., x^{(2n)}),$$

then system (2.2) has a periodic solution.

Conclusions. The existence of periodic solutions of nonlinear autonomous ordinary differential equations of nth order is investigated. The method used is based on the Brouwer's degree theory using homotopy invariance a property of topological degree.

REFERENCES

2. J. Cronin, " *Fixed Points and Topological Degree in Nonlinear Analysis*", American Mathematical Society, Providence, (1964).

2. N.G. Lloyd, "*Degree Theory*," Cambridge University Press, Cambridge (1978).

7. B. Mehri and H.A. Emamirad, "On the Existence of Periodic Solutions For Autonomous Second-order Systems," Nonlinear Anal., Theory Math., Meth. and Appl., **3**(5)(1978)pp. 577-582.

7. B. Mehri and M.A. Niksirat, "On the Existence of Periodic Oscillation for Vector Nonlinear Necond-Order System," in: Proceedings of the 28th Annual Iranian Mathematics Conference, 1997.

7. B. Mehri and M.A. Niksirat, "On the Existence of Periodic Solutions for the Quasi-Linear Third Order ODE," J. Math. Anal. Appl. **261**(1)(2001) pp. 159-167.

7. B. Mehri and M.A. Niksirat, "The Existence of Periodic Solutions For the Nonlinear Autonomous ODEs," Nonlinear Anal. Forum **5**(2000) pp. 163-171.

7. B. Mehri and M.A. Niksirat, "Some Computable Results on the Existence of Periodic Solutions for Singular Non-Autonomous Third Order Systems," Appl. Math. and Comp. **163**(2005) pp. 51-60.

On the Nonmonotone Behavior of the Newton-Gmback Method

Emil Cătinaş

Romanian Academy, "T. Popoviciu" Institute of Numerical Analysis, P.O. Box 68–1, Cluj-Napoca, Romania, e-mail: ecatinas@ictp.acad.ro

Abstract. GMBACK is a Krylov solver for large linear systems, which is based on backward error minimization properties. The minimum backward error is guaranteed (in exact arithmetic) to decrease when the subspace dimension is increased. In this paper we consider two test problems which lead to nonlinear systems which we solve by the Newton-GMBACK. We notice that in floating point arithmetic the mentioned property does not longer hold; this leads to nonmonotone behavior of the errors, as reported in a previous paper. We also propose a remedy, which solves this drawback.

Keywords: linear/nonlinear systems, Krylov solvers, Newton method.
PACS: 02.60.Cb

NONLINEAR SYSTEMS.

The Newton method for solving a nonlinear system $F(x) = 0$, $F : D \subseteq \mathbb{R}^N \to \mathbb{R}^N$ leads to the solving of a linear system at each iteration step:

$$F'(x_k) s_k = -F(x_k)$$
$$x_{k+1} = x_k + s_k, \qquad k = 0, 1, \ldots, \quad x_0 \in D.$$

Under the following conditions (which will be implicitly assumed throughout this paper) the Newton method converge locally at q-superlinear rate (see [7]):
- $\exists x^* \in \operatorname{int} D$ such that $F(x^*) = 0$
- the mapping F is Fréchet differentiable on $\operatorname{int} D$, with F' continuous at x^*;
- the Jacobian $F'(x^*)$ is invertible.

When one considers approximate Jacobians at each step, $F'(x_k) + \Delta_k \in \mathbb{R}^{N \times N}$, we are lead to the quasi-Newton (QN) iterates

$$(F'(x_k) + \Delta_k) s_k = -F(x_k)$$
$$x_{k+1} = x_k + s_k, \qquad k = 0, 1, \ldots, \quad x_0 \in D.$$

We have characterized the superlinear convergence of these iterates in the following result:

Theorem 1. *[3] Assume that the* QN *iterates converge to* x^**. Then the convergence is superlinear if and only if*

$$\|\Delta_k s_k\| = o(\|F(x_k)\|), \quad as\ k \to \infty. \tag{1}$$

THE GMBACK METHOD

When the dimension N is large, the numerical solving of a linear system

$$Au = b, \quad A \in \mathbb{R}^{N \times N} \text{nonsingular}, b \in \mathbb{R}^N,$$

becomes a difficult task. The Krylov solvers are popular choices for accomplishing this task, since they may offer good approximations at low computational cost.

CP1046, *Selected Papers from ICNAAM 2007 and ICCMSE 2007*
edited by T. E. Simos, G. Maroulis, G. Psihoyios, and Ch. Tsitouras

We shall consider here the GMBACK solver introduced by Kasenally in [6]. For a given subspace dimension $m \in \{1, \ldots, N\}$ and an initial approximation $u_0 \in \mathbb{R}^N$ having the residual $r_0 = b - Au_0$, it finds $u_m^{GB} \in u_0 + \mathcal{K}_m = u_0 + \text{span}\{r_0, Ar_0, \ldots, A^{m-1}r_0\}$ by the following minimization property:

$$\left\| \Delta_m^{GB} \right\|_F = \min_{u_m \in u_0 + \mathcal{K}_m} \|\Delta_m\|_F \quad \text{w.r.t.} \ (A - \Delta_m)u_m = b.$$

Here $\|\cdot\|_F$ denotes the Frobenius norm of a matrix, $\|Z\|_F = tr(ZZ^t)^{1/2}$ while $\|\cdot\|_2$ will denote the Euclidean norm from $\mathbb{R}^N$ and its induced operator norm.

The following steps are performed for determining u_m^{GB}:

Arnoldi

- Let $r_0 = b - Ax_0$, $\beta = \|r_0\|_2$ and $v_1 = \frac{1}{\beta}r_0$;
- For $j = 1, \ldots, m$ do
$$h_{ij} = (Av_j, v_i), \quad \text{for } i = 1, \ldots, j$$
$$\hat{v}_{j+1} = A\hat{v}_j - \sum_{j=1}^{i} h_{ij}v_i$$
$$h_{j+1,j} = \left\| \hat{v}_{j+1} \right\|_2$$
$$v_{j+1} = \frac{1}{h_{j+1,j}} \hat{v}_{j+1}$$
- Form the Hessenberg matrix $\bar{H}_m \in \mathbb{R}^{(m+1) \times m}$ with the (possible) nonzero elements h_{ij} determined above, and the matrix $V_m \in \mathbb{R}^{N \times m}$ having as columns the vectors v_j: $V_m = [v_1 \ldots v_m]$.

GMBACK

- Let $\beta = \|r_0\|_2$,
$\hat{H}_m = [-\beta e_1 \quad \bar{H}_m] \in \mathbb{R}^{(m+1) \times (m+1)}$, $\hat{G}_m = [u_0 \quad V_m] \in \mathbb{R}^{N \times (m+1)}$,
$P = \hat{H}_m^t \hat{H}_m \in \mathbb{R}^{(m+1) \times (m+1)}$ and $Q = \hat{G}_m^t \hat{G}_m \in \mathbb{R}^{(m+1) \times (m+1)}$;
- Determine an eigenvector v_{m+1} corresponding to the smallest eigenvalue λ_{m+1}^{GB} of the generalized eigenproblem $Pv = \lambda Qv$;
- If the first component $v_{m+1}^{(1)}$ is nonzero, compute the vector $y_m^{GB} \in \mathbb{R}^m$ by scaling v_{m+1} such that

$$\begin{bmatrix} 1 \\ y_m^{GB} \end{bmatrix} = \frac{1}{v_{m+1}^{(1)}} v_{m+1};$$

- Set $u_m^{GB} = x_0 + V_m y_m^{GB}$.

We shall assume in the following analysis that u_m^{GB} exists (this may not be the case when all the eigenvectors of the smallest eigenvalue have the first component 0).

Kasenally proved that for any $u_0 \in \mathbb{R}^N$ and $m \in \{1, \ldots, N\}$, the backward error Δ_m^{GB} corresponding to the GMBACK solution satisfies

$$\left\| \Delta_m^{GB} \right\|_F = \sqrt{\lambda_{m+1}^{GB}}. \tag{2}$$

Regarding the induced operator Euclidean norm, the following inequality is known:

$$\|Z\|_2 \leq \|Z\|_F, \quad \text{for all } Z \in \mathbb{R}^{N \times N}. \tag{3}$$

The eigenvalues at steps m and $m+1$ in the Arnoldi algorithm interlace as follows.

Lemma 2. *[6] Suppose that $m+1$ steps of the Arnoldi process have been taken and $h_{m+2,m+1} \neq 0$. Furthermore, assume that $\{\lambda_i^{GB}\}_{i=1,\ldots,m+1}$ and $\{\hat{\lambda}_i^{GB}\}_{i=1,\ldots,m+2}$ are, respectively, the eigenvalues of the matrix pairs (P_m, Q_m) and (P_{m+1}, Q_{m+1}) arranged in decreasing order. Then, for any $i \leq m+1$,*

$$\hat{\lambda}_i^{GB} \leq \lambda_i^{GB} \leq \hat{\lambda}_{i+1}^{GB}.$$

THE CONVERGENCE OF THE NEWTON-GMBACK METHOD

The Newton-GMBACK iterates may be written as

$$\left(F'(y_k) - \Delta_k^{GB}\right) s_k^{GB} = -F(y_k),\tag{4}$$

and we may control the convergence of the iterates by Theorem 1 with the aid of the backward error. It is worth noting that we may use formulas (2) and (3) to evaluate the magnitude of the backward error in the Euclidean norm.

We obtain:

Theorem 3. *Assume that the Newton-GMBACK iterates are well defined and converge to* x^*. *If*

$$\lambda_k^{GB} \to 0, \quad as\ k \to \infty,\tag{5}$$

then they converge superlinearly.

Of course, we may also use the inexact Newton model, and control the convergence of the iterates with the aid of residuals, see [4].

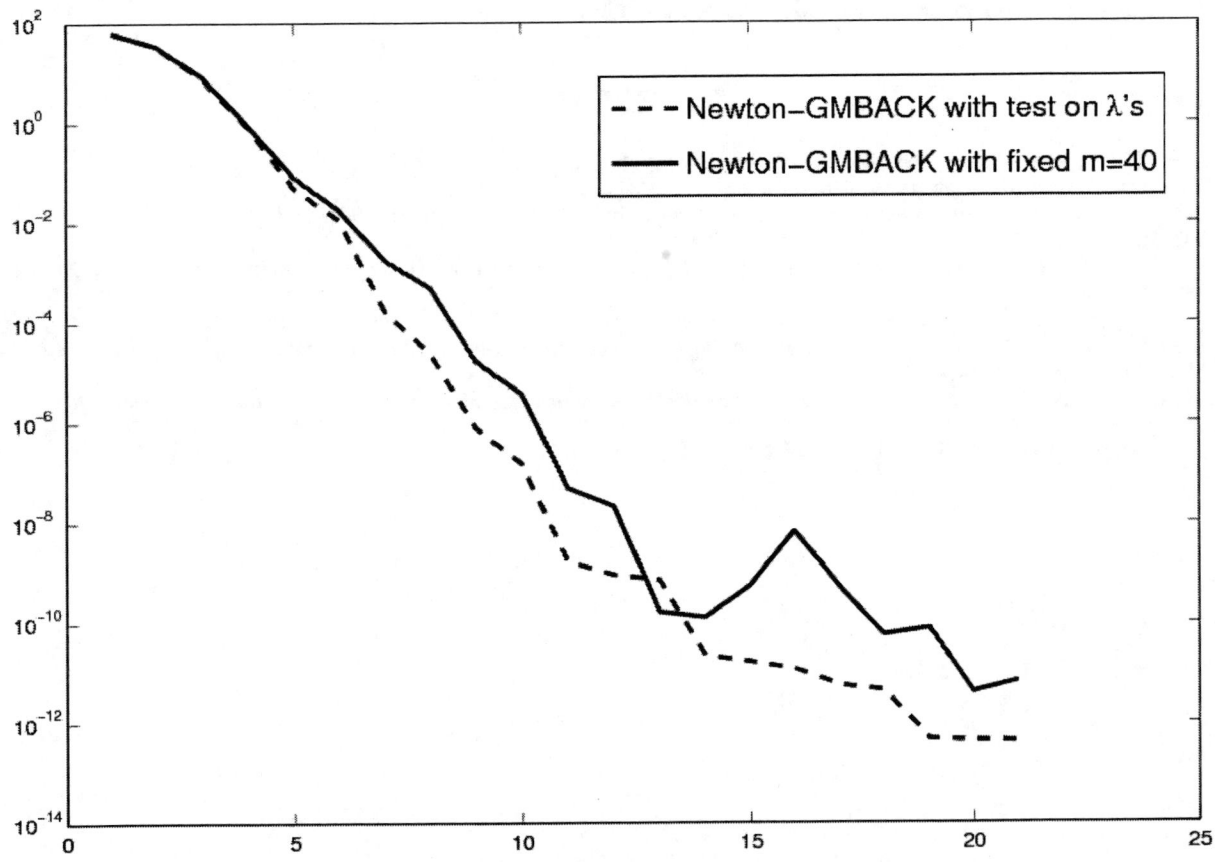

FIGURE 1.   Newton-GMBACK errors for the Bratu problem.

Bratu problem

Consider the nonlinear partial differential equation

$$-\triangle u + \alpha u_x + \lambda e^u = f,$$

over the unit square of $\mathbb{R}^2$, with Dirichlet boundary conditions. As mentioned in [1], this is a standard problem, a simplified form of which is known as the Bratu problem. We have discretized by 5-point finite differences, respectively by central finite differences on a uniform mesh, obtaining a system of nonlinear equations of size $N = (n-2)^2$, where n is the number of mesh points in each direction. As in [1], we took f such that the solution of the discretized problem to be the constant unity, and $\alpha = 10$, $\lambda = 1$, the initial approximations in the inner iterations were zero. The runs were made on a HP Proliant 570 G4 server, using MATLAB 2007a.

We took $N = 16,384$ and considered first some standard iterations, with fixed subspace dimension, of size $m = 40$. We noticed that, starting from outer iteration $k = 9$, where $\|F\|$ was of magnitude 1e−9, lemma 2 did not hold in floating point arithmetic. As we can see in figure 1, the consequence is that the convergence of the errors of iterates is no longer monotone.

The remedy we propose is to check at each inner iteration step in GMBACK whether the size of λ is decreasing, and to stop the iterations if the size is increasing. We obtain monotone behavior of the errors, and the plot in figure 1 is relevant for the runs we made.

ACKNOWLEDGMENTS

This research has been supported by MEdC under grant 2CEx06-11-96.

REFERENCES

1. P. N., Brown, and Y., Saad, *Hybrid Krylov Methods for Nonlinear Systems of Equations*, SIAM Journal on Scientific and Statistical Computing **11**, pp. 450–481 (1990).
2. E. Cătinaş, *Inexact perturbed Newton methods and applications to a class of Krylov solvers*, J. Optim. Theory Appl. **108**, pp. 543–570 (2001).
3. E. Cătinaş, *The inexact, inexact perturbed and quasi-Newton methods are equivalent models*, Math. Comp., 74 no. 249, pp. 291–301 (2005).
4. R. S. Dembo, S. C. Eisenstat and T. Steihaug, *Inexact Newton methods*, SIAM J. Numer. Anal. **19**, pp. 400–408 (1982).
5. J. E. Dennis, Jr. and J. J. Moré, *A characterization of superlinear convergence and its application to quasi-Newton methods*, Math. Comp. **28**, pp. 549–560 (1974).
6. E. M. Kasenally, *GMBACK: a generalised minimum backward error algorithm for nonsymmetric linear systems*, SIAM J. Sci. Comput. **16**, pp. 698–719 (1995).
7. J. M. Ortega and W. C. Rheinboldt, *Iterative Solution of Nonlinear Equations in Several Variables*, Academic Press, New York, 1970.

Boid Based Timetable Generation

Piotr Fulmański

Faculty of Mathematics and Computer Science University of Łódź, Banacha 22, Łódź 90-232

Abstract. The general timetabling problem consists of scheduling a number of events into a finite number of periods so that no individual entity is required to attend two or more events simultaneously. In most cases there will also be restrictions on the amount of resources, time etc. In this document we try to investigate a new method of timetable generation based not on popular evolutionary approach but on boids concept.

Keywords: Timetabling, boids.
PACS: 02.70

TIMETABLING PROBLEM

Timetabling problem is that we are looking for such assignment person–room–class–group–day to

- meet requirements of class considering limitations of rooms (number of seats, if there are any computers or projector etc.);
- some classes cannot be planed at the same time, because:
    - it is impossible for one group to have at the same time two or more different classes;
    - one person cannot have at the same time two or more different classes;
- some classes cannot be planed at the time which is not suitable for the tutor, e.g. because of person's others duties.

The simplicity of such formulated task is only apparently. In fact the timetabling problem is one of the hardest computational tasks and ranks to group of NP-Hard problems. Most of all proposed methods are based, more or less, on some kind of randomization (simulated annealing, evolutionary algorithms etc.). Thanks to it, it is possible to search all the domain in acceptable time[1]. Let us notice in this place, that if we want to use for example genetic algorithm, we have to create big initial population. For this task, this means, that we have to generate a great number of more or less acceptable timetables. Creation of initial population is by itself difficult task and also demands a lot of attention (e.g. [2]).

BOIDS

In 1986 year Craig Reynolds made a computer model of coordinated animal motion such as bird flocks and fish schools ([1]). The objects which map in digital world real creatures such as birds or fish were called boids. The rules of boids movement were very simple:

- Each boid determines its own movement direction in such a way that new direction will be more or less average of movement directions of boids from closest neighbourhood. Such behaviour seams to be justified – also people, when are in crowd and all crowd have the same aim, move like their closest neighbours (e.g. tourist trip, group of peoples runing to metro).
- Boids tend not to cross some minimal distance to their closest neighbours.
- Boids tend to stay in the middle of the group which consist of the closest neighbours.

---

[1] Of course by saying „all the domain" we mean not all the points from the domain but the representative points which belong to significantly different subdomains of the domain.

CP1046, *Selected Papers from ICNAAM 2007 and ICCMSE 2007*
edited by T. E. Simos, G. Maroulis, G. Psihoyios, and Ch. Tsitouras
© 2008 American Institute of Physics 978-0-7354-0574-5/08/$23.00

Each behaviour of the boid which does not break the above rules is acceptable. Local rules described in this way allow simulate really natural coordinated animal motion. So complicated process of global analyses of complex relations between all, sometimes numerous, objects we replace with locally controlled simple relations.

What is more interesting, by dedicating relatively little time, it is possible to model other group's behaviour, e.g. boids' dispersion after meeting an obstacle and boids' merging into one group after passing an obstacle. In this case it is enought to add a rule stating, that each boid tends not to cross some minimal distance to obstacle.

IDEA

Conception presented in previous section allows us to look at timetabling problem from local point of view. We consider the task not from the point of persons which are trying to prepare the timetable but from the point of persons for which this timetable is prepared.

It helps considerably, because we do not have to control all relations between objects (e.g. rooms, persons, groups etc.). There are lot of relations and necessity of its almost continuous checking is significant time consuming task.

Consider each person as equivalent with one boid, we can settle following rules:

- Each boid is in state *ready* if it has got some classes not connected with a room.
- Each boid which is in state *ready* chooses (randomly or according to some algorithm) one of its own classes not connected with a room. After that boids are in state *active*.
- Boids which are in state *active* choose (randomly or according to some algorithm) day in which they want to have classes. Let us notice that in this approach we do not have a problem with control if there is assignment person–day they did not want to — none of the boids acts to harm itself.
- Each boid which is in state *active* and chose a day, now chooses a room which meets requirement related to the resources required for specific classes. At this moment boid is being added to queue related to chosen room and is coming into state *stop*.
- Each queue periodically serves (according to some algorithm, e.g. FIFO) persons connected with it.
- Each boid which is being served is coming into state *reservation*. At this moment we can almost freely modify the behaviour of the boid. We can allow for reserving time for only one class or for as many classes as it is only possible. Once again we do not have problem with control if there is an assignment in hours the person did not want to (as above: none of the boids acting to harm itself).
- Boids which were served, and if still have some classes not connected with a room, are coming into state *ready*. In opposite case are coming into state *finish*.

If we try to compare such an approach with something, we can say that it resembles managing a company by chaos, which, surprisingly, in many cases gives quite good results. In real world timetabling in this way means that one day we collect all the people which have some classes and we say: „OK, go and reserve rooms for your classes". Before the end of the day we should obtain perfect timetable.

Saying more serious. Instead of building sophisticated procedures to control dependencies and take decision about assignment person–room–class–group–day considering all information, we can limit ourself to quite simple, short and fast procedures to control the most basic behaviours. Incorporating some random conditions, we ensure, similarly like e.g. in Monte-Carlo methods, that paying not to much we can search substantial part of the domain. It can happen that we will be unable to find optimal solution. Instead of it, we will be able to generate a lot of acceptable solutions in short time what sometimes is enough.

Implementation – main loop

Main loop of the approach implemented above is very simple:

```
while(true)
{
    person=GetPerson();
    if(person!=null)
    {
```

```
        person.ChooseSubject();
        person.ChooseDay();
        person.ChooseRoom();
    }

    ServePersons(HowMany.FIRST);

    if(StopConditions()==true)
        break;
    }
}
```

As we can see, there are only few, almost selfdescribing, functions.

- *GetPerson()* This function chooses one of the boids which is in *ready* state. Changing only this one function we can easily make that e.g. we will plan first of all classes for persons which have more classes than other. It is enough to write simply sorting algorithm. If among people with many classes we prefer professors we can add the second criteria to the sorting algorithm to adapt program to our expectations. No other changes will be essential.
- *ChooseSubject()* Function called for *person* object and selects classes for which in the next step we will be looking for a room. Like with *GetPerson()* also in this case we can easily modify the way of timetable construction. It is not so hard to variantize it in the sense of making different decisions depending on *person* representing real different person. For different person it can be differnet priority of classes to serve as a first.
- *ChooseDay(), ChooseRoom()* These functions choose, respectively, day in which we will be looking for a room and finally a room for choosen classes. When we choose a room we have to remember about requirements about resources (chairs, computers etc.).
- *ServePersons(HowMany.FIRST)* This function checks all of the queues and serves persons in these queues. The number of served persons depends on argument (in this example only the first person will be served and all next will have to wait for the next time).
- *StopConditions()* This function check general, easy to find, stop conditions: no empty room, no person with noassigned class etc.

TEST

Until now there were only some tests done with small amount of data to check correctness of idea and implemented algorithm. We assume that:

- Classes can be planed from Monday to Friday and from 8:00am to 8:00pm.
- There are three, independent, groups of students. Each group consist of two subgroups of students. So we have got: *group I, subgroup 1, group I, subgroup 2, group II, subgroup 1*, etc.
- The subgroups (30 students each) are constant — students cannot migrate from one to another.
-
- For each group we planned three classes called lectures which are obligatory for all students (two subgroups). So we have got: *Lect I 1* for first lecture for first group of students, *Lect II 3* for third lecture for second group of student etc.
- For each group we planned three classes called laboratories. There are two subgroups for each laboratory. So we have got: *Lab III 1 2* for first laboratory for second subgroup for third group of students etc.
- From above we can see that at the same time it cannot be for the same group: two or more lectures, lecture and any of laboratories and laboratories for the same subgroup.
- There are three teachers. All of them have some classes from each group.
- The teachers do not want to have classes from 8:00am to 12:00am and one of them additionally does not want to have classes from 4:00pm to 6:00pm.
- There are four rooms with 30 seats and one with 60 seats.

Every time we test, we obtain different timetable which fulfills the conditions presented above.

SUMMARY

Proposed approach does not provide optimal solution. Instead of it, we can generate many admissible solutions in very short time. It is likely that some of them, when corrected by human, can be perceived by teachers and students as optimal.

In this way we can also initialize population for genetic algorithm.

Performed tests are promising and let us expect that this approach turn out effective also for great number of data.

REFERENCES

1. Craig Reynolds, Boid, http://www.red3d.com/cwr/boids/.
2. E. K. Burke, J. P. Newall, R. F. Weare, *Initialization Strategies and Diversity in Evolutionary Timetabling*, Evolutionary Computation 6(1), 1998 Massachusetts Institute fo Tehnology, pp. 81-103.
3. A.R. Mushi, *Mathematical programming formulations for the examinations timetable problem: the case of the University of Dar El Salaam*, African Journal of Science and Technology, Science and Engineering Series Vol. 5, No. 2, pp 34-40

Perspective Pose Estimation with Geometric Algebra

Christian Gebken and Gerald Sommer

Institute of Computer Science, Chair of Cognitive Systems, Christian-Albrechts-University of Kiel, Germany

Abstract. A novel method which entirely resides inside conformal geometric algebra (CGA) is presented estimating the pose of a camera from one image of a known object. At first, subproblems covering only three feature points are solved and globally assessed. The object model is accordingly pruned and rigidly fitted to corresponding projection rays by evaluating a succinct CGA expression which emerged from a purely geometric approach. It results a set of 3-point poses each given by a motor. These spinor elements of CGA embody rigid body motions from the manifold $SE(3)$. The poses are then to be averaged according to their quality. This is the second aspect of this work as the respective motors do not come from a linear space and averaging must be carried out appropriately. For this purpose, a technique called weighted intrinsic mean is used.

Keywords: geometric (Clifford) algebra, conformal space, geometric calculus, perspective projection, intrinsic mean, Lie groups
PACS: 02.10.Xm, 02.20.Sv, 02.40.Dr

INTRODUCTION

In pose estimation the orientation and position of an internally calibrated camera is recovered from its images. For this purpose, the 3D point-model of at least one pictured object is assumed to be known together with a set of correspondences, which interrelate model points and image points. This kind of pose estimation is often referred to as 'perspective n-point problem' (PnP).

The classic but challenging task of pose estimation is from the field of computer vision. Most approaches to that subject are iterative, highly nonlinear or require an initialization. Closed form solutions to the 3-point problem (P3P), where the number of correspondences is three, exist [1] but may result in up to four distinct solutions because P3P is not necessarily unique. As extension to P3P it is also possible to consider four points. Fischler and Bolles [2], for example, take subsets and perform consistency checks to eliminate the P3P ambiguity for most point configurations. In [3] Quan and Lan present an algorithm capable of finding the unique solution to PnP. They first generate a global system of linear equations based on all correspondences. Next, the exact 3D-vectors to the object points w.r.t. the camera coordinate system are estimated. Finally, camera orientation and position are evaluated one after another. But this class of techniques is shown in [4] to improperly model the physical imaging, i.e. a perspective projection must be considered. Rosenhahn and Sommer [5] formulate algebraic constraints with CGA. They obtain a hybrid system of linear equations based on correspondences between points, lines and between point and line. Starting from an initialization the pose is iteratively estimated in 3D. It is to mention that such global PnP approaches are not able to spot and disregard false or noisy correspondences.

In this text we derive a vivid geometric formulation of P3P with CGA. At the same time, this motivates a sound selection strategy for point triplets, i.e. not all possible 3-combinations in the correspondences must be considered. Solutions of PnP are rigid body motions (RBM) from the manifold $SE(3)$. We show that the respective P3P-solution is fully determined by a certain angle θ^*. Our geometric approach further leads to an algebraic function $h(\theta) \in \mathbb{R}$, with θ^* being a root of which. For each root the corresponding RBM is globally assessed regarding its effect on the entire n-point scenario. The set of 3-point candidate solutions can then be reduced by solutions from obviously false correspondences. The remaining RBMs, at most one for every triplet considered, are finally averaged by means of the weighted intrinsic mean, which is tailored to elements of $SE(3)$.

Geometry with Geometric Algebra

For a detailed introduction to geometric algebra (GA) see e.g. [6, 7]. Here we only convey a minimal framework. We consider the geometric algebra $\mathbb{G}_{4,1} = \mathscr{C}\ell(\mathbb{R}^{4,1}) \supset \mathbb{R}^{4,1}$ of the 5D conformal space (CGA), cf. [8, 9]. Let $\{\mathbf{e}_1, \mathbf{e}_2, \mathbf{e}_3, \mathbf{e}_+, \mathbf{e}_-\}$ denote the basis of the underlying Minkowski space $\mathbb{R}^{4,1}$, where $\mathbf{e}_1^2 = \mathbf{e}_2^2 = \mathbf{e}_3^2 = \mathbf{e}_+^2 = 1 = -\mathbf{e}_-^2$. We

CP1046, *Selected Papers from ICNAAM 2007 and ICCMSE 2007*
edited by T. E. Simos, G. Maroulis, G. Psihoyios, and Ch. Tsitouras

use the common abbreviations $\mathbf{e}_\infty = \mathbf{e}_- + \mathbf{e}_+$ and $\mathbf{e}_o = \frac{1}{2}(\mathbf{e}_- - \mathbf{e}_+)$ for the point at infinity and the origin, respectively. The algebra elements are termed multivectors, which we symbolize by capital bold letters, e.g. A. A juxtaposition of algebra elements, like CR, indicates their geometric product, being the associative, distributive and noncommutative algebra product. We denote '·' and '∧' the inner and outer (exterior) product, respectively. A point $x \in \mathbb{R}^3 \subset \mathbb{G}_{4,1}$ of the 3D Euclidian space is mapped to a conformal point (null vector) X, with $X^2 = 0$, by $X = x + \frac{1}{2}x^2\mathbf{e}_\infty + \mathbf{e}_o$. A geometric object, say O, can now be defined in terms of its inner product null space $\mathbb{X}(O) = \{X \in \mathbb{R}^{4,1} | X \cdot O = 0\}$. This implies an invariance to scalar multiples $\mathbb{X}(O) = \mathbb{X}(\lambda O)$, $\lambda \in \mathbb{R} \backslash \{0\}$. Verify that $S = s + \frac{1}{2}(s^2 - r^2)\mathbf{e}_\infty + \mathbf{e}_o$ represents a 2-sphere centered at $s \in \mathbb{R}^3$ with radius r. Similarly, if $n \in \mathbb{R}^3$, with $n^2 = 1$, then $P = n + d\,\mathbf{e}_\infty$ represents a plane at a distance d from the origin with orientation n. Using this vector valued geometric primitives higher order entities can be build: given two objects, e.g. the planes P_1 and P_2, their line of intersection L is simply the outer product $L = P_1 \wedge P_2$. An important GA operation is the reflection. The reflection of A in O is given by the sandwiching product $B = OAO$. The most general case is the reflection in a sphere, which actually represents an inversion. Note that any RBM can be represented by consecutive reflections in planes. In CGA the resulting elements, for example $M = P_1 P_2$, are referred to as motors, cf. [5]. Some object C would then be subjected to $MC\tilde{M}$, whereby the symbol '~' stands for an order reversion, i.e. the reverse of M is $\tilde{M} = P_2 P_1$. Since reflection is an involution, a double reflection $O(OAO)O$ must be the identity w.r.t $\mathbb{X}(A)$, therefore $O^2 \in \mathbb{R}$ by associativity. It looks like a conclusion, but in GA every vector valued element $X \in \mathbb{R}^{4,1}$ squares to a scalar by definition $X^2 := X \cdot X \in \mathbb{R}$. Using the above definitions, we have $S^2 = r^2$ and $P^2 = 1$.

THALES' THEOREM REVISITED

In this section we demonstrate how a simple geometric theorem motivates a solution to P3P. Our considerations refer to the left and middle part of figure 1.

The generalization of Thales' theorem states that, given a circle K, the centric angle $\angle(x_1', m, x_1)$ at m is twice the peripheral angle $\angle(x_1', O, x_1)$ at O. We use this fact and define two successive rotations: the first rotates x_1 to x_1' and the second rotates x_1' back onto the straight line connecting O and x_1. We obtain point x_1''. It is crucial, that any second point x_2 on K will also move towards O. Moreover, the distance from x_1 to x_2 stays constant since rotations are distance preserving.

Before we enlighten the value of this observation we work out the general transformation, that we denote R_θ, in terms of CGA. We therefore replace each rotation with two reflections in suitable planes. We have to take care that the dihedral angle between the planes equals half of the rotation angle. Further, the planes' line of intersection must coincide with the rotation axis. The two rotations can ultimately be realized by four reflections in the three planes P_1, P_2 and P_3. The order of application must be P_1, P_2, P_3 and P_1 again. We obtain the motor $R_\theta = P_1 P_3 P_2 P_1$. For the derivation we take a canonical coordinate system as a basis. The whole setup is depicted in figure 1. We define $P_1 = \mathbf{e}_1$, $P_2 = \cos(\theta)\mathbf{e}_1 + \sin(\theta)(\mathbf{e}_2 + r\mathbf{e}_\infty)$ and $P_3 = \cos(\theta/2)\mathbf{e}_1 + \sin(\theta/2)\mathbf{e}_2$, whereby r denotes the radius of circle K. After some algebra we obtain

$$R_\theta = \cos(\theta/2) + \sin(\theta/2)\underbrace{\left[\mathbf{e}_1\mathbf{e}_2 + r\left((\cos(\theta) + 1)\mathbf{e}_1\mathbf{e}_\infty + \sin(\theta)\mathbf{e}_2\mathbf{e}_\infty\right)\right]}_{L_\theta} = \exp\left(\theta/2\,L_\theta\right). \quad (1)$$

The element L_θ is a line representing the rotation axis of R_θ and plays the role of the imaginary unit i of complex numbers, for $L_\theta^2 = -1$. By noting that $R_\theta = P_1(P_3 P_2)P_1$ it can be recognized that R_θ is the reflection of $R_\theta' := P_3 P_2$ in plane P_1. Hence R_θ rotates by an angle θ being twice the dihedral angle between P_3 and P_2.

Regarding pose estimation this result can be interpreted as a way to rigidly move two 3D model points on their respective projection rays. The latter can be computed from the internal calibration and the corresponding image points. The point O represents the optical center of the camera. Unfortunately, it is not sufficient to consider two points, but the above result leads to a 3-point approach.

THE PERSPECTIVE 3-POINT PROBLEM

In this section, we concentrate on the subproblem of determining the pose of point triplets. We thus consider three 3D model points $\{x_1, x_2, x_3\}$ and their corresponding image points. From these we compute the respective projection rays

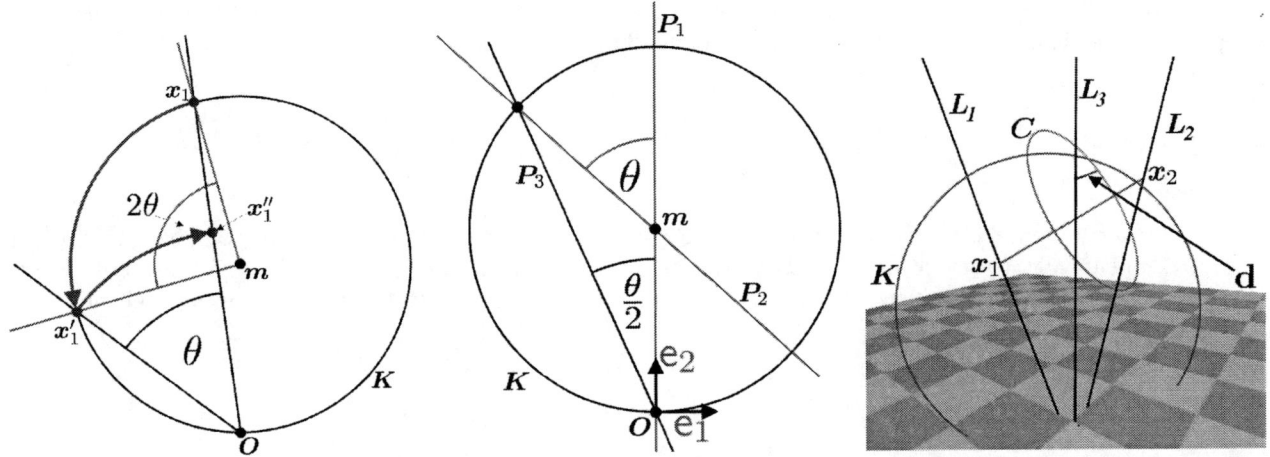

FIGURE 1. **Left:** the generalization of Thales' theorem explains why two successive rotations form a translation. **Middle:** the transformation can be realized by a sequence of four reflections in three well-chosen planes P_1, P_2 and P_3. The overall transformation is then $R_\theta = P_1 P_3 P_2 P_1$. **Right:** Distance **d** between circle C_θ and the third projection ray L_3.

$\{L_1, L_2, L_3\}$.

The aim is to find a position and orientation for the model such that each model point coincides with its respective projection ray. This problem is referred to as fitting. It enables the computation of an RBM that interrelates the external (world) coordinate system (WCS) of the 3D model and the camera coordinate system (CCS). Note that the RMB, which we encode in a motor, constitutes the camera pose.

The solution arises from the question where the third point x_3 might be while we use R_θ to move x_1 and x_2 along their projection rays. Clearly the mutual distances between the three points must me retained. The locus of the third point must therefore be a circle C which has to be subjected to R_θ along with x_1 and x_2, see figure 1. Note[1] that we have reached our aim if $C_\theta = R_\theta C \widetilde{R_\theta}$ intersects ray L_3. The CGA expression $C_\theta \wedge L$ represents a sphere which degenerates to the point of intersection in case L_3 hits C_θ. Hence a suitable function is $h(\theta) = (C_\theta \wedge L)^2 \in \mathbb{R}$ which has to attain zero.

Since the derivatives of $h(\theta)$ are analytically available, we employ the Newton-Raphson method for the computation of the at most four roots $\Theta = \{\theta_1, \theta_2, \ldots, \theta_k\}$, $k \leq 4$. The application of the respective motor $R_{\theta \in \Theta}$ to the points $\{x_1, x_2, x_3\}$ yields the fitted points $\{x'_1, x'_2, x'_3\}$, being the connection between CCS and WCS. For each angle $\theta \in \Theta$ we estimate the interrelating motor M_θ by means of standard methods, see [10]. The application of the $M_{\theta \in \Theta}$ to the entire n-point problem assesses their quality and ultimately reveals inappropriate pose candidates.

It remains to specify a selection strategy for triplets in order to put a limit on the computational complexity $\binom{n}{3}$: we select those triplets the image points of which form maximum area triangles on the image plane. In this way the impact of noise in the coordinates of the image points is minimized, i.e. the fitting is more constrained as the noise induced jittering of the model does not carry much weight.

THE PERSPECTIVE n-POINT PROBLEM

The issue regarding the fusion of the P3P motors $\{M_1, M_2, \ldots, M_N\}$ is the second key aspect of this work. Since any motor is from the Lie group $SE(3)$, being a manifold, the customary arithmetic mean may not be used: $A, B \in SE(3) \nRightarrow A + B \in SE(3)$. The Lie group $SE(3)$ is connected to its Lie algebra $se(3)$ (tangent space to the identity element of $SE(3)$) by the exp/log map. Note that in $se(3)$ any customary mean can be built as the algebra elements form a vector space. This is exploited by the 'weighted intrinsic mean', in which the log map is used to compute first-order mean approximations via the tangent space, see [11, 12]. The N motors are input to the outlined algorithm below, whereby

---

[1] The three original model points were initially subjected to an RBM such that x_1 and x_2 of the resulting points $\{x_1, x_2, x_3\}$ already lie on their corresponding projection rays. We then use $\{x_1, x_2, x_3\}$ to determine K and $C = C_0$.

the weights w_i, $1 \leq i \leq N$, reflect the motor assessments. Starting from the motor $\boldsymbol{M} = \mathtt{identity}$ the subsequent three steps are repeated until $\|\mathtt{log}(\Delta \boldsymbol{M})\|$ falls below a certain threshold ε.

$$
\begin{aligned}
1. && \Delta \boldsymbol{A}_i &= \log\left(\boldsymbol{M}^{-1}\boldsymbol{M}_i\right) && 1 \leq i \leq N \\
2. && \Delta \boldsymbol{M} &= \exp\left(\tfrac{1}{W}\sum_{i=1}^{N} w_i \Delta \boldsymbol{A}_i\right) && W = \sum_{i=1}^{N} w_i \\
3. && \boldsymbol{M} &= \boldsymbol{M}\Delta \boldsymbol{M}
\end{aligned}
\tag{2}
$$

Notice that the motor $\boldsymbol{M}$ is repeatedly updated by the residuals $\Delta \boldsymbol{M}$, which originate from the weighted averaging of algebra elements $\Delta \boldsymbol{A}_i$, $1 \leq i \leq N$. The term $\boldsymbol{M}^{-1}\boldsymbol{M}_i$ in step 1 moves the input closer to the identity element of $SE(3)$ in order to minimize the averaging error in step 2. A derivation of the algorithm and a uniqueness proof is given in [13].

Here we do not report experimental results but state that our pose estimation has already been vastly used. Besides, in synthetic experiments the method has proven to produce results competitive to the ground truth or to a stochastically optimal solution.

CONCLUSION

Starting from a clear geometric concept, we have proposed a solution to the perspective n-point problem which is free from initialization. Our method is consistent since it always yields the correct solution in the absence of noise. The involved 3-point problem has been streamlined to a scalar valued function solely depending on an angle. Our approach enables the rejection of solution candidates which stem from false correspondences. We have introduced a selection strategy for point triplets which reduces the computational effort and improves the accuracy at the same time. In the final weighted averaging of rigid body motions we have taken their algebraic nature into account using the intrinsic mean. Evidently, this work is also a valuable contribution to the field of (conformal) geometric algebra, which turned out to be the ideal framework to deal with geometric objects and transformations.

REFERENCES

1. R. M. Haralick, C.-N. Lee, K. Ottenberg, and M. Noelle, "Review and Analysis of Solutions of the Three Point Perspective Pose Estimation Problem," in *International Journal on Computer Vision*, 1994, vol. 13, pp. 331–356.
2. M. A. Fischler, and R. C. Bolles, "Random Sample Consensus: A Paradigm for Model Fitting with Applications to Image Analysis and Automated Cartography," in *Commun. ACM*, ACM Press, New York, NY, USA, 1981, vol. 24, pp. 381–395.
3. L. Quan, and Z.-D. Lan, "Linear N-Point Camera Pose Determination," in *PAMI*, 1999, vol. 21, pp. 774–780.
4. B. K. P. Horn, Projective Geometry Considered Harmful (1999), available at http://www.ai.mit.edu/people/bkph/papers/harumful.pdf.
5. B. Rosenhahn, and G. Sommer, "Pose estimation in conformal geometric algebra, Part I: the stratification of mathematical spaces," in *Journal of Mathematical Imaging and Vision*, Springer Science + Business Media, Inc., 2005, vol. 22, pp. 27–48.
6. D. Hestenes, and G. Sobczyk, *Clifford Algebra to Geometric Calculus: A Unified Language for Mathematics and Physics*, Reidel, Dordrecht, 1984.
7. L. Dorst, D. Fontijne, and S. Mann, *Geometric Algebra for Computer Science: An Object-Oriented Approach to Geometry*, Morgan Kaufmann Publishers Inc., San Francisco, CA, USA, 2007.
8. P. Angles, "Construction de revêtements du groupe conforme d'un espace vectoriel muni d'une «métrique» de type (p,q)," in *Annales de l'institut Henri Poincaré*, 1980, vol. 33, pp. 33–51.
9. H. Li, D. Hestenes, and A. Rockwood, "Generalized homogeneous coordinates for computational geometry," in *Geometric computing with Clifford algebras: theoretical foundations and applications in computer vision and robotics*, Springer-Verlag, London, UK, 2001, pp. 27–59.
10. K. S. Arun, T. S. Huang, and S. D. Blostein, "Least-Squares Fitting of Two 3-D Point Sets," in *PAMI*, 1987, vol. 9, pp. 698–700.
11. P. T. Fletcher, C. Lu, and S. C. Joshi, "Statistics of Shape via Principal Geodesic Analysis on Lie Groups," in *CVPR*, 2003, pp. 95–101.
12. S. Buchholz, and G. Sommer, "On Averaging in Clifford Groups," in *Computer Algebra and Geometric Algebra with Applications*, edited by H. Li, P. Olver, and G. Sommer, 6th International Workshop IWMM 2004, Springer, 2005, vol. 3519 of *LNCS*, pp. 229–238.
13. S. R. Buss, and J. P. Fillmore, "Spherical averages and applications to spherical splines and interpolation," in *Commun. ACM*, ACM Press, New York, NY, USA, 2001, vol. 20, pp. 95–126.

On the Hyers-Ulam Stability of a bi-Jensen Functional Equation

Gwang-Hui Kim*, Yang-Hi Lee† and Dal-Won Park**

*Department of Mathematics, Kangnam University, Suwon 449-702, Republic of Korea
†Department of Mathematics Education, Gongju National University of Education
of Education, Gongju 314-060, Republic of Korea
**Department of Mathematics Education, Kongju National University, Gongju, 314-701, Republic of Korea

Abstract. In this paper, we prove the stability of a bi-Jensen functional equation

$$4f(\frac{x+y}{2}, \frac{z+w}{2}) = f(x,z) + f(x,w) + f(y,z) + f(y,w).$$

Keywords: bi-Jensen functional equation
PACS: Primary 39B52

INTRODUCTION

Throughout this paper, let X and Y be vector spaces. A mapping $g : X \to Y$ is called a Cauchy mapping (respectively a Jensen mapping) if g satisfies the functional equation $g(x+y) = g(x) + g(y)$ (respectively $2g(\frac{x+y}{2}) = g(x) + g(y)$)[2,4,6]. For a given mapping $f : X \times X \to Y$, we define

$$Df(x,y,z,w) := f(x+y, z+w) - f(x,z) - f(x,w) - f(y,z) - f(y,w),$$

$$D_1 f(x,y,z,w) := 2f(x+y, \frac{z+w}{2}) - f(x,z) - f(x,w) - f(y,z) - f(y,w),$$

$$D_2 f(x,y,z,w) := 2f(\frac{x+y}{2}, z+w) - f(x,z) - f(x,w) - f(y,z) - f(y,w),$$

$$D_3 f(x,y,z,w) := 4f(\frac{x+y}{2}, \frac{z+w}{2}) - f(x,z) - f(x,w) - f(y,z) - f(y,w)$$

for all $x,y,z,w \in X$. A mapping $f : X \times X \to Y$ is called a biadditive (Cauchy-Jensen, Jensen-Cauchy, bi-Jensen, respectively) mapping if f satisfies the functional equations $Df = 0(D_1 f = 0, D_2 f = 0, D_3 f = 0$, respectively).

In 2006, Park and Bae obtained the generalized Hyers-Ulam stability of a Cauchy-Jensen mapping[4] and the generalized Hyers-Ulam stability of a bi-Jensen mapping [1]. In 2007, Jun and Lee [3] improved the Park and Bae's results.

In this paper, we prove the Hyers-Ulam stability of a Cauchy-Jensen functional equation and the Hyers-Ulam stability of a bi-Jensen functional equation.

THE STABILITY OF A CAUCHY-JENSEN FUNCTIONAL EQUATION

In this section, let Y be a Banach space. We need the following lemma to prove Theorem 2.2.

Lemma 2.1 *Let $A = \{x \in X : \|x\| \le 1\}$. If $F : X \times X \to Y$ satisfies the equality*

$$D_1 F(x,y,z,w) = 0$$

for all $x,y,z,w \in X \backslash A$, then there exists a unique Cauchy-Jensen map $F' : X \times X \to Y$ satisfying the equality

$$F'(x,y) = F(x,y)$$

CP1046, *Selected Papers from ICNAAM 2007 and ICCMSE 2007*
edited by T. E. Simos, G. Maroulis, G. Psihoyios, and Ch. Tsitouras
© 2008 American Institute of Physics 978-0-7354-0574-5/08/$23.00

for all $x, y \in X \backslash A$.

Theorem 2.2 *Let $f : X \times X \longrightarrow Y$ be a mapping satisfying*

$$\|D_1 f(x, y, z, w)\| \leq \varepsilon$$

for all $x, y, z, w \in X \backslash A$. Then there exists a unique Cauchy-Jensen mapping $F : X \times X \longrightarrow Y$ satisfying

$$\|f(x, y) - F(x, y)\| \leq \frac{\varepsilon}{2}$$

for all $x, y \in X \backslash A$. In particular, the map F is given by

$$F(x, y) = \lim_{n \to \infty} \frac{f(2^n x, y)}{2^n}$$

for all $x, y \in X \backslash A$.

THE STABILITY OF A BI-JENSEN FUNCTIONAL EQUATION

In this section, let Y be a Banach space. We need the following lemma(Lemma 1 in [3]) to prove Theorem 3.2.

Lemma 3.1 *Let $f : X \times X \longrightarrow Y$ be a bi-Jensen mapping. Then*

$$f(x, y) = \frac{1}{2^n} f(2^n x, y) + \frac{1}{2^n}(1 - \frac{1}{2^n}) f(0, 2^n y) + (1 - \frac{1}{2^n})^2 f(0, 0)$$

for all $x, y \in X$ and $n \in N$.

Theorem 3.2 *Let $f : X \times X \longrightarrow Y$ be a mapping satisfying*

$$\|D_3 f(x, y, z, w)\| \leq \varepsilon$$

for all $x, y, z, w \in X$. Then there exists a unique bi-Jensen mapping $F : X \times X \longrightarrow Y$ satisfying

$$\|f(x, y) - F(x, y)\| \leq \varepsilon$$

for all $x, y \in X$ with $F(0, 0) = f(0, 0)$. In particular, the map F is given by

$$F(x, y) = \lim_{n \to \infty} \frac{f(2^n x, y) + f(0, 2^n y)}{2^n} + f(0, 0)$$

for all $x, y \in X$.

Lemma 3.3 *Let $A = \{x \in X : \|x\| \leq 1\}$. If $F : X \times X \longrightarrow Y$ satisfies the equality*

$$D_3 F(x, y, z, w) = 0$$

for all $x, y, z, w \in X \backslash A$, then there exists a unique bi-Jensen map $F' : X \times X \longrightarrow Y$ satisfying the equality

$$F'(x, y) = F(x, y)$$

for all $x, y \in X \backslash A$.

Theorem 3.4 *Let $f : X \times X \longrightarrow Y$ be a mapping satisfying*

$$\|D_3 f(x, y, z, w)\| \leq \varepsilon$$

for all $x, y, z, w \in X \backslash A$. Then there exists a unique bi-Jensen mapping $F : X \times X \longrightarrow Y$ satisfying

$$\|f(x, y) - F(x, y)\| \leq \frac{7\varepsilon}{4}$$

for all $x, y \in X \backslash A$ with $F(0, 0) = f(0, 0)$. In particular, the map F is given by

$$F(x, y) = \lim_{n \to \infty} \left(\frac{f(2^n x, y)}{2^n} + \frac{f(0, 2^n y)}{2^n}\right) + f(0, 0)$$

for all $x, y \in X \backslash A$.

REFERENCES

1. *J.-H.Bae and W.-G. Park* On the solution of a bi-Jensen functional equation and its stability, *Bull. Korean Math. Soc. (2006)* 499-507

2. *D.H. Hyers,* On the stability of the linear functional equation, *Pro. Nat'l. Acad. Sci. U.S.A.* **27** *(1941)222–224.*

3. *K.-W. Jun and Y.-H. Lee,* On the Hyers-Ulam-Rassias stability of a Cauchy-Jensen functional equation, *to appear.*

4. *Th.M. Rassias,* On the stability of the linear mapping in Banach spaces, *Proc. Amer. Math. Soc. 72 (1978)297–300.*

5. *W.-G. Park and J.-H. Bae,* On a Cauchy-Jensen functional equation and its stability, *J. Math. Anal. Appl.* **323** *(2006)634–643.*

6. *S. M. Ulam* Problems in Modern Mathematics *Chap. VI, Science eds., Wiley, Newyork, 1960.*

Towards Analytic Solutions of Step-Wise Safe Switching for Known Affine-Linear Models

Fotis N. Koumboulis and Maria P. Tzamtzi

Department of Automation, Halkis Institute of Technology, 34400 Psahna, Evia, Greece

Abstract. In the present work we establish conditions which guarantee safe transitions for the closed-loop system produced by the application of the Step-Wise Safe Switching control approach to an affine linear system when the nonlinear description of the plant is known. These conditions are based on the local Input to State Stability (ISS) properties of the nonlinear system around the plant's nominal operating points.

Keywords: Switching Control, Nonlinear Systems, Input-to-State Stability, Lyapunov Functions
PACS: 02.30.Yy, 07.05.Dz, 89.20.Bb, 89.20.Kk

INTRODUCTION

Logic-based switching supervisory control (see f.e. [1]-[4]) has been proposed to control nonlinear plants, that cannot be adequately controlled by a single field controller. Thus, a supervisory controller is designed to orchestrate switching between a set of candidate controllers, as the plant's I/O trajectories move between different areas of operation. This approach is particularly suited for controlling nonlinear plants with linear controllers.

Step-Wise Safe Switching (SWSS), introduced in [4] for the case of single input-single output (SISO) nonlinear systems with unknown description, is based on experimental data and rules, aiming to provide a framework that achieves safe transitions between operating points of a nonlinear plant. SWSS performs switching between a set of controllers that achieve specific performance requirements simultaneously for more than one linearizations of the plant corresponding to adjacent nominal operating points. However, the presentation of SWSS in [4] does not provide the conditions under which the proposed switching algorithm guarantees safe transitions.

In the present work we deal with this problem for the case when the plant is described by a known first-order affine linear model. The derived conditions are based on the local Input-to-State-Stability (ISS) of the corresponding nonlinear model around the nominal operating points of the plant.

SAFE SWITCHING CONDITIONS

Model of the Plant

Consider the first order SISO affine linear system described by:

$$\dot{y} = f(y) + g(y)u \tag{1}$$

where $y \in \square$ is the output of the system, $u \in \square$ is the input of the system and $f, g \in C^1$. Consider also that $g(y) \neq 0$ for all y or at least for all y within the plant's operating range under consideration. Let $[Y_0, U_0]$ denote an operating point of the nonlinear system (1) determined by the equation

$$f(Y_0) + g(Y_0)U_0 = 0 \tag{2}$$

The plant's behavior may be locally approximated, around the operating point $[Y_0, U_0]$, by the linearization:

$$\Delta \dot{y} = a(U_0, Y_0)\Delta y + g(Y_0)\Delta u \tag{3}$$

CP1046, *Selected Papers from ICNAAM 2007 and ICCMSE 2007*
edited by T. E. Simos, G. Maroulis, G. Psihoyios, and Ch. Tsitouras
© 2008 American Institute of Physics 978-0-7354-0574-5/08/$23.00

where the variables Δy and Δu are approximations of $\delta y = y - Y_0$ and $\delta u = u - Y_0$, respectively, while $a(U_0, Y_0) = (df(y)/dy)_{y=Y_0} + (dg(y)/dy)_{y=Y_0} U_0$.

SWSS Algorithm

Let $L = \{\ell_1, \ell_2, ..., \ell_\kappa\}$, with $\ell_i = [Y_i, U_i]$, denote a set of nominal operating points of plant (1), and $S_i, i = 1, ..., \kappa$ denote a set of linear models that describe approximately the nonlinear plant's behavior in a neighborhood of the corresponding nominal operating points ℓ_i. The main characteristics of the Step-Wise Safe Switching algorithm are summarized in the following [4]:

a)   For each nominal operating point ℓ_i, we determine corresponding operating areas, inside which the lineal model S_i constitutes a satisfactory approximation of the nonlinear system (1).

b)   The set of nominal operating points is selected dense enough, so that the operating areas of adjacent operating points overlap (*dense web principle* [4])

c)   For each pair (ℓ_i, ℓ_{i+1}) of adjacent operating points, a common controller $C_{i,i+1}$ is designed, that satisfies a set of desired design requirements simultaneously for both linear models S_i and S_{i+1}.

d)   Transition always takes place between operating points lying in the same or within adjacent operating areas. During transition from any operating point in O_{i-1} to any operating point in O_i, the controller $C_{i-1,i}$ is applied [4]. Before initiating transition to a third operating point in O_{i+1}, switching to controller $C_{i,i+1}$ takes place.

e)   Controller switching is allowed to take place only when the I/O trajectories are close to an operating point.

When the plant's nonlinear description is unknown, the linear models, as well as the corresponding operating areas, are determined using exclusively experimental data [see [4]).

Areas of Local ISS

The following theorem establishes local ISS ([5]-[7]) for the nonlinear plant's trajectories around the nominal operating points and determines the corresponding areas of stability in the (u, y) plane.

Theorem 1: Let $[Y_i, U_i]$ denote a nominal operating point of the nonlinear system (1). Assume that $a(U_i, Y_i) < 0$, with $a(U_i, Y_i)$ as determined in (3). Then, the nonlinear system (1), expressed with respect to the variable δy:

$$\delta \dot{y} = f(\delta y + Y_i) + g(\delta y + Y_i)U_i + g(\delta y + Y_i)\delta u \qquad (4)$$

is $(\varepsilon_i, \delta_i)$-ISS ([5]-[7]). More specifically, there exists $r_i > 0, \lambda_i > 0$ and $k_i > 0$, such that for each $|\delta y(t_0)| < \delta_i = r_i$ and for all $\delta u(t)$ such that $\sup_{t \geq t_0} |\delta u(t)| < \varepsilon_i = r_i / \lambda_i$, it holds that

$$|\delta y(t)| \leq |\delta y(t_0)| e^{-k_i(t-t_0)} + \lambda_i \sup_{t_0 \leq \tau \leq t} |\delta u(\tau)|, \quad \forall t \geq t_0 \qquad (5)$$

Moreover, a selection of r_i, λ_i and k_i that satisfies (5) is given by the following: $k_i = -[a(U_i, Y_i) + \mu_i + g_b(r_i)/\lambda_i]$, where μ_i, r_i and λ_i are strictly positive real numbers and $g_b(r_i)$ is a continuous increasing function of r_i that satisfy the conditions

$$|h(\delta y, U_i, Y_i)| < \mu_i, \forall |\delta y| < r_i \qquad (6)$$

$$|g(\delta y + Y_i)| \leq g_b(r_i), \forall |\delta y| \leq r_i, \ g_b(r_i) \geq |g_b(0)| = |g(Y_i)| \qquad (7)$$

$$a(U_i, Y_i) + \mu_i + g_b(r_i)/\lambda_i < 0 \qquad (8)$$

where $h(\delta y, U_i, Y_i) = -a(U_i, Y_i) + d[f(\delta y + Y_i) + g(\delta y + Y_i)U_i]/d(\delta y)|_{\delta y = z \in [0, \delta y]}$.

Proof: The linearization of the nonlinear system (4) around its zero equilibrium point is given by (3). Since (3) is asymptotically stable, then taking into account that $f, g \in C^1$, the nonlinear system (4) is locally asymptotically stable and consequently $(\varepsilon_i, \delta_i)$-ISS ([5]-[7]). To determine the parameters ε_i and δ_i, consider the Lyapunov function $V(\delta y) = (\delta y)^2 / 2$. Using the mean value theorem, it holds that $f(\delta y + Y_i) + g(\delta y + Y_i)U_i = a(U_i, Y_i)\delta y +$

$h(\delta y, U_i, Y_i)\delta y$, where $h(\delta y, U_i, Y_i) \to 0$ as $\delta y \to 0$. Hence for each $\mu_i > 0$ there exists $r_i > 0$ such that (6) holds. The continuity of $g(y)$ implies that there exists a function $g_b(r_i)$ such that (7) holds. Finally, $a(U_i, Y_i) < 0$ implies that by selecting a sufficiently small μ_i and a sufficiently large λ_i, condition (8) is satisfied. Combining the above statements, it follows that for the time derivative of the Lyapunov function $V(\delta y)$ along the trajectories of the nonlinear system (4), it holds that for all $|\delta y| < r_i$ and $|\delta y| \ge \lambda_i |\delta u|$:

$$\dot{V}(\delta y) = \delta y \delta \dot{y} = \delta y(f(\delta y + Y_i) + g(\delta y + Y_i)U_i + g(\delta y + Y_i)\delta u) \le -k_i(\delta y)^2 \qquad (9)$$

Hence, according to [5], [6], condition (5) holds for the nonlinear system (4). ∎

Safe Transitions

Based on the local ISS properties of the nonlinear system, sufficient conditions for safe transitions, are established in the following, for the closed-loop system derived by the application of the SWSS algorithm.

Theorem 2: Consider the affine linear system (1) and let $L = \{\ell_1, \ell_2, \ldots, \ell_\kappa\}$ a set of nominal operating points for this plant. Assume that $a(U_i, Y_i) < 0$ for all $\ell_i \in L$. Let $R_i = \{(u, y) : |y - Y_i| < \delta_i, |u - U_i| < \varepsilon_i\}$, $i = 1, \ldots, \kappa$ define a rectangles around ℓ_i in the (u, y) plane, inside which the corresponding nonlinear systems (8) are $(\varepsilon_i, \delta_i)$-ISS. Then, the application of the SWSS algorithm guarantees safe transitions provided that:

   i) $R_i \cap R_{i+1} \ne \varnothing, i = 1, \ldots, \kappa - 1$ (10)

ii) The part of the operating curve $u = -f(y)/g(y)$, connecting the nominal operating points $\ell_1, \ell_2, \ldots, \ell_\kappa$ lies within the union $R_1 \cup \cdots \cup R_\kappa$.

iii) All desired transitions take place between operating points belonging within the rectangle R_i of at least one nominal operating point $\ell_i \in L$.

iv) For all transitions initiated from an operating point lying within R_i, the applied controller results in values of the control variable within the range determined by R_i, that is $|u(t) - U_i| < \varepsilon_i$ for all $t \ge 0$.

Proof: According to the local ISS properties of the nonlinear system (1) established in Theorem 1, the plant's I/O trajectories remain within a known bounded neighborhood of the operating point ℓ_i provided that a) $|y(0-) - Y_i| < \delta_i$ and b) $|u(t) - U_i| < \varepsilon_i$ for all $t \ge 0$. Moreover, it can be easily verified from (5), that in this case the I/O trajectories either lie within R_i for sufficiently large values of time, or at least tend asymptotically to the boundaries of R_i as $t \to \infty$. According to the SWSS algorithm all step transitions are initiated when the plant's I/O trajectories are close to an operating point. Thus, condition (iii) implies that all step transitions of the SWSS algorithm satisfy condition (a) for at least one nominal operating point $\ell_i \in L$. Conditions (i)-(iii) imply that for all initial operating point ρ_s in R_i and any desired final operating point ρ_f in R_{i+1}, it is possible to design an I/O trajectory, that starts from ρ_s and ends at ρ_f, while always remaining sufficiently close either to ℓ_i or ℓ_{i+1}. Finally, condition (iv) guarantees that the common controllers applied by the SWSS algorithm, indeed succeed to keep the I/O trajectories sufficiently close to at least one nominal operating point, thus guaranteeing safe transitions. ∎

CASE STUDY

Consider the process of an isothermal continuous stirred tank reactor (CSTR) ([8], [9]), in which an irreversible reaction $A \to B$ takes place. The isothermal CSTR is described by the first order affine linear differential equation:

$$\frac{dy(t)}{dt} = -\rho y(t) + \frac{(\alpha - y(t))}{V} u(t) \qquad (11)$$

where the output y of the CSTR process is the concentration of the reactant A in the reactor, the input u of the process is the flow of the incoming and outgoing liquid, $V = 2000[l]$ denotes the constant volume of the liquid within the tank, $\alpha = 1.0[\text{mol/l}]$ denotes the concentration of the reactant A in the incoming liquid and $\rho = 0.028[\text{min}^{-1}]$ denotes the rate constant of the reaction. The reaction temperature is assumed to be constant.

Based on the constructive characteristics of the reactor, the control variable u is known to be bounded according to the inequality $0[l/\text{min}] \leq u(t) \leq 504[l/\text{min}]$. The operating range of the output value that will be considered in the present study is given by $0[\text{mol}/l] \leq y \leq 0.9[\text{mol}/l]$.

Consider the following three nominal operating points of the CSTR: $\ell_1 = [Y_1, U_1] = [0.45, 45.81]$, $\ell_2 = [Y_2, U_2] = [0.545, 67.08]$ and $\ell_3 = [Y_3, U_3] = [0.605, 85.77]$. The corresponding linearized models around each operating point $[Y_i, U_i]$ are given by $d\Delta y(t)/dt = -(\rho + U_i/V)\Delta y(t) + (\alpha - Y_i)\Delta u(t)/V$. Since U_i is positive, it follows that all linearizations are asymptotically stable, with $a(U_i, Y_i) = -(\rho + U_i/V) < 0$. The nonlinear system (4) corresponding to each operating point ℓ_i is given by

$$\delta\dot{y} = -\rho(\delta y + Y_i) + (a - \delta y - Y_i)(\delta u + U_i)/V = -(\rho + U_i/V)\delta y + (a - \delta y - Y_i)\delta u/V \tag{12}$$

For the derivative of the Lyapunov function $V(\delta y) = (\delta y)^2/2$ along the trajectories of system (12) it holds that:

$$\dot{V}(\delta y) = -(\delta y)^2(\rho + U_i/V) + (a - \delta y - Y_i)\delta u \delta y/V \leq -(\delta y)^2\left[\rho + U_i/V - (a + r - Y_i)/(\lambda V)\right] \tag{13}$$

for all $|\delta y| < r_i = \min\{Y_i, 0.9 - Y_i\}$ and $|\delta y| \geq \lambda_i|\delta u|$, where $\lambda_i > 0$ is appropriately selected so that $k = \rho + U_i/V - (a + r_i - Y_i)/(\lambda_i V) > 0$. This implies that for each operating point ℓ_i, system (12) is $(\varepsilon_i, \delta_i)$-ISS, with $\delta_i = r_i$ and $\varepsilon_i = \delta_i/\lambda_i$, thus defining the following areas of local ISS:

$$R_1 = \{(u, y) : |y - 0.45| < 0.45, |u - 45.81| < 45\}$$

$$R_2 = \{(u, y) : |y - 0.545| < 0.355, |u - 67.08| < 50.71\}$$

$$R_3 = \{(u, y) : |y - 0.605| < 0.295, |u - 85.77| < 59\}$$

It is obvious that the rectangles R_1, R_2 and R_3 satisfy condition (i) of Theorem 2. Moreover, it can be readily verified that the part of the operating curve of the CSTR connecting the operating points $[0.014, 0.81]$ and $[0.72, 144.77]$ lies within the union $R_1 \cup R_2 \cup R_3$. Then according to Theorem 2, all transitions between operating points lying on this part of the operating curve are safe provided that the applied controller satisfies condition (iv) of Theorem 2.

Note that the CSTR process has also been used as a case study in [4], for the application of the SWSS algorithm, considering the plant's nonlinear model to be unknown, and designing the SWSS algorithm based exclusively on experimental data. We note that the conditions imposed by Theorem 2 are slightly stricter, compared with those of [4], with respect to the range of values for the control variable u during the transient stage of a step transition. This is due to the fact that the upper bounds with respect to u of the tolerance operating areas for ℓ_1, ℓ_2 and ℓ_3 are larger than the corresponding of R_1, R_2 and R_3.

ACKNOWLEDGMENTS

The present work is co-financed by the Hellenic Ministry of Education and Religious Affairs' and the ESF of the European Union within the framework of the "Operational Programme for Education and Initial Vocational Training" (Operation "Archimedes-I").

REFERENCES

1. A. Leonessa, W.M. Haddad, V. Chelaboina,, *IEEE Trans. on Autom. Control* **46**, 17-28 (2001).
2. M.W. McConley, B.D. Appleby, M.A. Dalheh, E. Feron, *IEEE Trans. on Autom. Control* **45**, 33-49 (2000).
3. E.F. Costa, V.A. Oliveira, *Automatica* **38**, 1247-1250 (2002).
4. F.N. Koumboulis, R.E. King, A. Stathaki, *Information Sciences* **177**, 2736-2755 (2007).
5. A. Teel, L. Praly, *SIAM J. Control Optim.* **33**, 1443-1488 (1995).
6. H.K. Khalil, *Nonlinear Systems*, Macmillan Publishing Company, New York, 1996
7. E.D. Sontag, *IEEE Trans. Autom. Control* **34**, 435-443 (1989).
8. F.J. Doyle III, H.S. Kwatra, J.S. Schwaber, *Chemical Eng. Science* **53**, 2675-2690 (1998).
9. M.A. Henson, D.E. Seborg, *A.I.Ch.E. J.,* **36**, 1753–1757 (1990).

A Mathematical Model for the Middle Ear Ventilation

G. Molnárka[1], E. M. Miletics[2] and M. Fücsek[3]

[2]*Széchenyi István University, Győr, Hungary*
[3]*Petz Aladár County Teaching Hospital, Győr, Hungary*
[1]*molnarka@sze.hu,* [2]*miletics@sze.hu,* [3]*drfucsek@t-online.hu*

Abstract. The otitis media is one of the mostly existing illness for the children, therefore investigation of the human middle ear ventilation is an actual problem. In earlier investigations both experimental and theoretical approach one can find in ([1]-[3]). Here we give a new mathematical and computer model to simulate this ventilation process. This model able to describe the diffusion and flow processes simultaneously, therefore it gives more precise results than earlier models did. The article contains the mathematical model and some results of the simulation.

Keywords: computer simulation, ear ventilation.
PACS: N 87.10.Ed

THE PROBLEM FOR SOLVING

It is well known that the middle ear pressure must be kept equal the atmospheric pressure level for keeping a normal middle ear function. Sometimes for some reasons the middle ear ventilation through the Eustachian tube can break off, therefore the normal function of the ear will be hurt. In order to investigate this phenomenon different attempts have been made from measuring the middle ear pressure to building different physical and mathematical models. The earlier mathematical models are relatively rough, therefore we attempt to construct and investigate a mathematical – computer model describing the problem. To do this first we have to formulate a geometrical model of the middle ear.

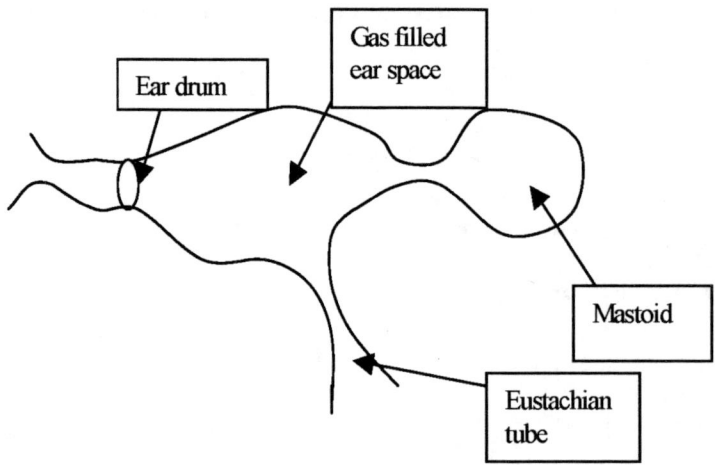

FIGURE 1. The geometrical scheme of the middle ear

In normal case the ventilation takes place through the Eustachian tube in each minute. Between two ventilation processes the tube is closed. During this time pressure changes within the closed ear space and this process governed by gas diffusion through the surrounding tissue. The main gas components are in the middle ear volume: nitrogen, oxygen, carbon dioxide and water-vapour.

CP1046, *Selected Papers from ICNAAM 2007 and ICCMSE 2007*
edited by T. E. Simos, G. Maroulis, G. Psihoyios, and Ch. Tsitouras
© 2008 American Institute of Physics 978-0-7354-0574-5/08/$23.00

MODEL FOR THE DIFFUSION PROCESSES IN MIDDLE EAR SPACE

The process of gas diffusion can be described by the Noyes-Whitney equations. This process takes place in a closed part of middle ear, therefore the diffusion of gas from the tissue and into the tissue are closely connected processes. Therefore its accurate mathematical model for general case is a coupled system of differential equations. If we assume, that the coupling effects are not essential in description of the main effects, neglecting these components we can deduce the following system of differential equations: (see [4])

$$\frac{dP_{O_2}(t)}{dt} = -\frac{1}{2.4 \cdot 10^7} \frac{P_t}{V_0} A \cdot \tilde{D}_{O_2} (P_{O_2}(t) - P_{O_2(tissue)}(t)) \tag{1}$$

$$\frac{dP_{N_2}(t)}{dt} = -\frac{1}{2.4 \cdot 10^7} \frac{P_t}{V_0} A \cdot \tilde{D}_{N_2} (P_{N_2}(t) - P_{N_2(tissue)}(t)) \tag{2}$$

$$\frac{dP_{CO_2}(t)}{dt} = -\frac{1}{2.4 \cdot 10^7} \frac{P_t}{V_0} A \cdot \tilde{D}_{CO_2} (P_{CO_2}(t) - P_{CO_2(tissue)}(t)) \tag{3}$$

$$\frac{dP_{H_2O}(t)}{dt} = -\frac{1}{2.4 \cdot 10^7} \frac{P_t}{V_0} A \cdot \tilde{D}_{H_2O} (P_{H_2O}(t) - P_{H_2O(tissue)}(t)) \tag{4}$$

where we use the following notations: the functions $P_{O_2}(t)$, $P_{CO_2}(t)$, $P_{N_2}(t)$, $P_{H_2O}(t)$ are the partial pressures of nitrogen, oxygen, water and carbon dioxide gases respectively in the middle ear space, the quantities $P_{O_2(tissue)}(t)$, $P_{CO_2(tissue)}(t)$, $P_{N_2(tissue)}(t)$, $P_{H_2O(tissue)}(t)$ are the same in the tissue, $P_t = P_{O_2}(t) + P_{CO_2}(t) + P_{N_2}(t) + P_{H_2O}(t)$. One important approximation is the assumption, that $P_t = const$, so it is independent from time. By this assumption we decouple the system of equations. V_0 is the volume, A is the surface of the middle ear space, $\tilde{D}_{CO_2}, \tilde{D}_{O_2}, \tilde{D}_{N_2}, \tilde{D}_{H_2O}$ are the normalized diffusion coefficients of the gas components through tissue wall.

A normal human ear works as follows. During circ. 60 sec the Eustachian tube is closed, then the tube will be open during ab. 0.4 sec. Then this process periodically repeats. In the first part of the process it is controlled by gas diffusion while during the second part is very fast pressure equalization process between the middle ear space and atmosphere, and change of the concentration of the gas componens in the middle ear space takes place. The mathematical model of the processes during the first part is given by equations (1)-(4) if we formulate initial conditions for them. The initial conditions will be given by formula (11)-(12).

MODEL FOR PRESSURE EQUALIZATION AND MIXING GAS COMPONENTS IN THE MIDDLE EAR

By the mathematical models of pressure equalization thru the Eustachian tube the process is very fast. For the middle ear space the volume of the gas flowing in or out per sec is $X_t'(t)A$, where A is the cross-section of the Eustachian tube and $X(t)$ is the solution of the differential equations as follows:

$$\frac{d^2 X(t)}{dt^2} = \frac{1}{h\rho_b(0)}(P_t(0) - P_k(t)(1 + \frac{A}{V_0}(X(t) - X(0)))), \quad if \quad P_t(0) > P_k(t) \tag{5}$$

or

$$\frac{d^2 X(t)}{dt^2} = \frac{1}{h}(\frac{P_t(0)}{1 - \frac{A}{V_0}(X(t) - X(0))} - P_k(t))\frac{1}{\rho_k(t)}, \quad if \quad P_t(0) < P_k(t) \tag{6}$$

With initial conditions $X(0) = 0$ and $\frac{dX(t)}{dt}\big|_{t=0} = 0$. Here h means the length of the Eustachian tube, $P_k(t)$ is the pressure of the ambient atmosphere, ρ_b and ρ_k are the mass density of the gas mixture in the middle ear and ambient air respectively.

The investigation of the solutions of the equations (5)-(6) shows that the equalization of the pressure difference takes place through 0.001-0.1 sec depending on parameters in the equations. So our model confirms the assumption of ear specialists who assert, that 0.4 sec enough for the pressure equalization process.

MODEL FOR MIXING GAS COMPONENTS DURING PRESSURE EQUALIZATION PROCESS

The pressure equalization process means that gas mixture flows in or flows out to or from middle ear volume. Therefore the concentration of the gas mixture in the middle ear space changes. To describe the whole process we have to give the model of this changing process too. If the gas pressure equalization process begins with the case described by equation (5), the gas flows out from middle ear to air, therefore the concentration of the gas in the middle ear space does not change. In that case we have to calculate only the concentration of the gas filling the middle ear space at the end of the diffusion process. To do that the volume difference of the gas components at the end of the diffusion process are:

$$\Delta V_{CO_2} = \frac{\tilde{D}_{CO_2} A}{2.4 \cdot 10^7} \int_0^T (P_{CO_2}(t) - P0_{CO_2}) dt \qquad (7)$$

$$\Delta V_{O_2} = \frac{\tilde{D}_{O_2} A}{2.4 \cdot 10^7} \int_0^T (P_{O_2}(t) - P0_{O_2}) dt \qquad (8)$$

$$\Delta V_{N_2} = \frac{\tilde{D}_{N_2} A}{2.4 \cdot 10^7} \int_0^T (P_{N_2}(t) - P0_{N_2}) dt \qquad (9)$$

$$\Delta V_{H_2O} = \frac{\tilde{D}_{H_2O} A}{2.4 \cdot 10^7} \int_0^T (P_{H_2O}(t) - P0_{H_2O}) dt \qquad (10)$$

where $P0_{O_2}, P0_{N_2}, P0_{CO_2}, P0_{H_2O}$ are the partial pressure of the gas components in the tissue assumed to be constant, the functions $P_{N_2}(t), P_{O_2}(t), P_{CO_2}(t), P_{H_2O}(t)$ are the solution of the equations (1)-(4). Let $\Delta V = \Delta V_{N_2} + \Delta V_{O_2} + \Delta V_{CO_2} + \Delta V_{H_2O}$ and let $\tilde{V}_0 = V_0 + \Delta V$. T is the length of the time period while the ear is closed.

If $\tilde{V}_0 > V_0$, this is the case when the gas flows out from the middle ear to the ambient air. Then the concentration or partial pressure of gas components is at the end of the diffusion process:

$$P_{N_2,new} = \frac{V_{N_2}(t=0) + \Delta V_{N_2}}{\tilde{V}_0} P_k, \quad P_{CO_2,new} = \frac{V_{CO_2}(t=0) + \Delta V_{CO_2}}{\tilde{V}_0} P_k,$$

$$P_{H_2O,new} = \frac{V_{H_2O}(t=0) + \Delta V_{H_2O}}{\tilde{V}_0} P_k, \quad P_{O_2,new} = \frac{V_{O_2}(t=0) + \Delta V_{O_2}}{\tilde{V}_0} P_k, \qquad (11)$$

If $\tilde{V}_0 < V_0$, this is the case when air flows into the middle ear. Then the concentration or partial pressure of gas components is at the end of the diffusion process:

$$P_{N_2,new} = \frac{V_{N_2}(t=0) + \Delta V_{N_2} - \Delta V \dfrac{P_{k,N_2}}{P_k}}{V_0} P_k, \quad P_{CO_2,new} = \frac{V_{CO_2}(t=0) + \Delta V_{CO_2} - \Delta V \dfrac{P_{k,CO_2}}{P_k}}{V_0} P_k,$$

$$P_{H_2O,new} = \frac{V_{H_2O}(t=0) + \Delta V_{H_2O} - \Delta V \dfrac{P_{k,H_2O}}{P_k}}{V_0} P_k, \quad P_{O_2,new} = \frac{V_{O_2}(t=0) + \Delta V_{O_2} - \Delta V \dfrac{P_{k,O_2}}{P_k}}{V_0} P_k, \qquad (12)$$

Where $P_{k,N_2}, P_{k,O_2}, P_{k,H_2O}, P_{k,CO_2}$ are the partial pressures of the components in the ambient air.

NUMERICAL SOLUTION OF THE MODEL, RESULTS

The mathematical model for the middle ear ventilation given by the system of differential equation (1)-(4) and the initial conditions described by the formula (11)-(12). The particularity of this problem is that however the simplified differential equations can be solved easily, the initial conditions are time and solution dependent. Such problem can be solved only by numerical algorithm. To do this we worked out a symbolic-numeric algorithm and we realized it by using the MAPLE system because of the symbolic steps in the algorithm.

One can see a result of the solution on the Figure 2..

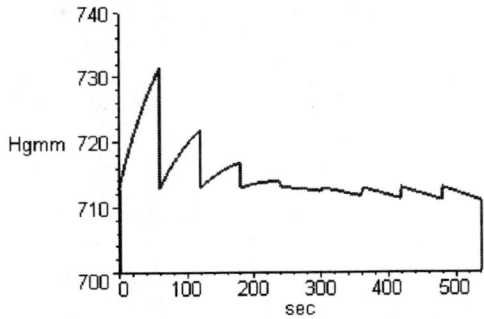

FIGURE 2. The pressure in the middle ear without the partial pressure of the vapour

One can see the result of the calculation for the middle ear pressure on the Figure 2. when we omit the change of the partial pressure of the water vapour. This is real approximation when the ventilation takes place thru the Eustachian tube. From the results we can see, that the process tends to a periodic change of the pressure in the middle ear and the value of the function describing the pressure is determined only by the geometrical parameter of the middle ear and does not depend on the initial partial pressures of the gas components. The time needed for the system to come near the steady periodic state is relatively small but to attain the steady periodic solution for the system needs approximately one day.

CONCLUSION

The mathematical model given is a system of differential equation coupled by periodically changing initial conditions. The computer algorithm given is suitable to calculate numerically the solution of the model. The results obtained are in accordance with the measurements but its study gives more detailed picture about the middle ear ventilation processes than it was known before.

ACKNOWLEDGMENTS

This was supported by the project OMFB-00567/2007.

REFERENCES

1. Ksel Grontved, Agot Moller,Lars Jörgensen, "Studies on Gas Tension in the Normal Middle Ear," in Acta Otolaryngol (Stockh) 1990; 109: pp. 271-277
2. Kathleen Chin, Orval E. Brown, Scott C. Manning, Carl C. Crandell, "Middle Ear Pressure Variation: Effect of Nitrous Oxide," The Laryngoscope 107, March 1997. pp. 357-362.
3. Mihály Fücsek, "Normal Middle Ear ventilation: a New Theory," Orvosi Hetilap, Akadémiai Kiadó, Vol. 148, N⁰ 22/June 2007, pp.1033-1035.
4. Győző Molnárka, Edit M.-Miletics, "A mathematical Model for Middle Ear Ventilation," Tech. Rep. 2007/6. Ental Ltd.

Numerical Analysis of Thermal Comfort at Open Air Spaces

K. Papakonstantinou[a], C. Belias[b], S. Pantos-Kikkos[c] and A. Assana[c]

[a]Department of Aircraft Technology, Technological Educational Institute of Chalkis, Psahna 34 400 Evia, Greece
[b]Department of Chemistry, University of Athens, Zografou, Panepistimiopolis, 157 71 Athens, Greece
[c]Studio 75 Architects, Demokratias Av., 154 51 Athens, Greece

Abstract. The present paper refers to the numerical simulation of air velocity at open air spaces and the conducting thermal comfort after the evaluation of the examined space using CFD methods, taking into account bioclimatic principles at the architectural design. More specially, the paper draws attention to the physical procedures governing air movement at an open environment area in Athens (urban park), named "Attiko Alsos", trying to form them in such way that will lead to the thermal comfort of the area's visitors. The study presents a mathematical model, implemented in a general computer code that can provide detailed information on velocity, prevailing in three-dimensional spaces of any geometrical complexity. Turbulent flow is simulated and buoyancy effects are taken into account. This modelling procedure is intended to contribute to the effort towards designing open areas, such as parks, squares or outdoor building environments, using thermal comfort criteria at the bioclimatic design. A computer model of this kind will provide the architects or the environmental engineers with powerful and economical means of evaluating alternative spaces' designs.

Keywords: Open Environment Design, CFD methods, Thermal Comfort, PMV, Bioclimatic Design.
PACS: 44.05.+e, 44.25.+f, 44.90.+c, 47.10.-g, 47.11.-j

INTRODUCTION

The study presents a CFD simulation that can provide detailed information concerning the environmental conditions of an examined area in Athens, named "Attiko Alsos". The analysis is accomplished at the framework of the project: "Recovery – Reconstruction of Park Attiko". The target of the analysis is to conduct the bioclimatic evaluation of the examined area, using computational fluid dynamics methods in order to propose an architectural space's design, based on bioclimatic principles, in order to succeed the thermal comfort of the users and the visitors of the place. The technique can be used with confidence at least for checking the relative advantages and disadvantages of various design alternatives in planning external urban spaces [1].

MATHEMATICAL FORMULATION AND APPLICATION OF THE MODEL

For steady flow, the equations for continuity and velocity components can be expressed in the following general conservation form for the general variable φ:

$$\frac{\partial}{\partial x_i}\left(\rho u_i \varphi\right) = \frac{\partial}{\partial x_i}\left(\Gamma_\varphi \frac{\partial \varphi}{\partial x_i}\right) + S_\varphi$$

(1)

where ρ is the density, u_i the velocity vector components, Γ_φ the effective exchange coefficient of φ and S_φ the source rate per unit volume.

CP1046, *Selected Papers from ICNAAM 2007 and ICCMSE 2007*
edited by T. E. Simos, G. Maroulis, G. Psihoyios, and Ch. Tsitouras
© 2008 American Institute of Physics 978-0-7354-0574-5/08/$23.00

The source rate and the effective exchange coefficient corresponding to each variable φ solved for in this study are given in Table 1. The magnitudes being referred in this table are: μ the viscosity, σ the Prandtl number for, k the kinetic energy of turbulence, ε the eddy dissipation rate and $G = \mu_1 (\partial u_i / \partial x_j + \partial u_j / \partial x_i) \partial u_i / \partial x_j$ the turbulence production rate. The values of the constants C1 and C2 are 1.44 and 1.92, respectively.

TABLE 1. Source rates and effective exchange coefficient of variable φ

Equation	φ	S_φ	Γ_φ
Continuity	1	0	0
Momentum	u_i	$-\dfrac{\partial p}{\partial x_i} + \rho_{ref} g_i$	μ
Kinetic energy of turbulence	k	$G - \rho\varepsilon$	μ/σ_k
Eddy dissipation rate	ε	$C_1 \dfrac{\varepsilon}{k} G - C_2 \rho \dfrac{\varepsilon^2}{k}$	μ/σ_ε

To solve the set of the model differential equations together with their boundary conditions, a finite-domain technique is used which combines features of the methods of Patankar and Spalding and a whole-field pressure-correction solver. The space dimensions are discretised into finite intervals and the variables are computed at only a finite number of locations, at the so-called "grid points". These variables are connected with each other by algebraic equations derived by their counterparts by integration over the control volumes defined by the above intervals. This leads to equations of the form:

$$\sum_n (A_n^\varphi + C)\varphi_P = \sum_n A_n^\varphi \varphi_n + CV$$

(2)

where the summation n is over the cells adjacent to a defined point P. The coefficients A_n^φ, which account for convective and diffusive fluxes across the elemental cell, are formulated using hybrid differencing. The source terms are written in the linear form $S_\varphi = C(V - \varphi)$ where C, V stand for a coefficient and a value of the variable φ.

Pressures are obtained from a pressure correction equation that yields the pressure change needed to procure velocity changes to satisfy the mass continuity. To solve the 3-D flow equations the "SIMPLEST" practice of Spalding is followed, in which the finite-domain coefficients of the momentum equations contain only diffusion contributions, the convection terms being added to the linearised source term. The momentum equations are solved by a point-by-point procedure. The pressure-correction equation is solved over the whole-field [2, 3].

TEST CASES CONSIDERED - RESULTS AND DISCUSSION

The examined place is presented in the topographical diagram at the next Figure 1. Two main areas will be examined, Area A and Area B. Two cases are simulated: Case 1 concerns the bioclimatic design of area A where a summer cinema, two tennis courts, a basketball court and a playground are situated. Case 2 concerns the bioclimatic design of area B where an open air amphitheatre, two tennis courts, a playground and the central square are situated. For both the cases wind velocity is considered 3.0 m/s, directed north-east, for both summer and winter season. Wind temperature at summertime is 27.4 °C. For the determination of the wind initial and boundary characteristics, the average maximum annual values are considered, as they prevail from existing meteorological data for the examined area, for a series of years.

Some indicatives results of the study are presented in the next two figures in the form of velocity vectors. Figure 2a presents the wind flow field for Area A and Figure 2b presents the wind flow field for Area B.

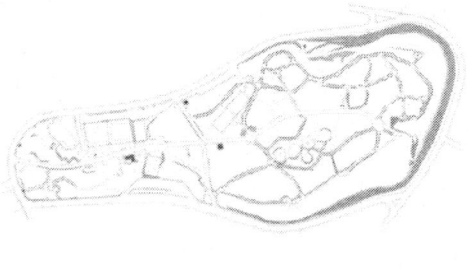

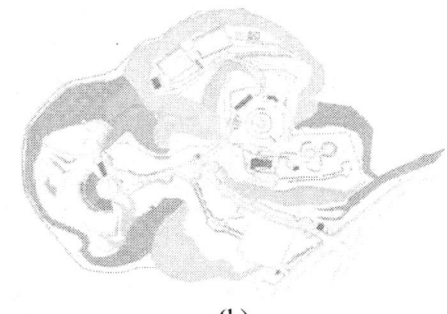

(a)

(b)

FIGURE 1. (a) Topographical diagram – Area A
(b) Topographical diagram – Area B

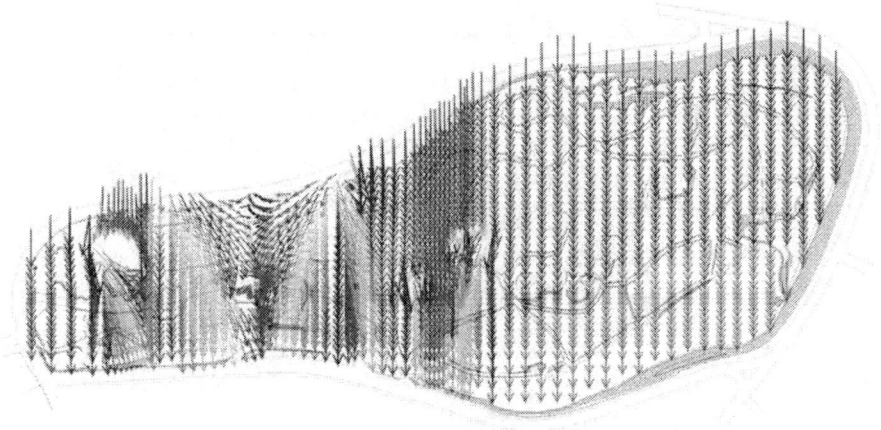

(a)

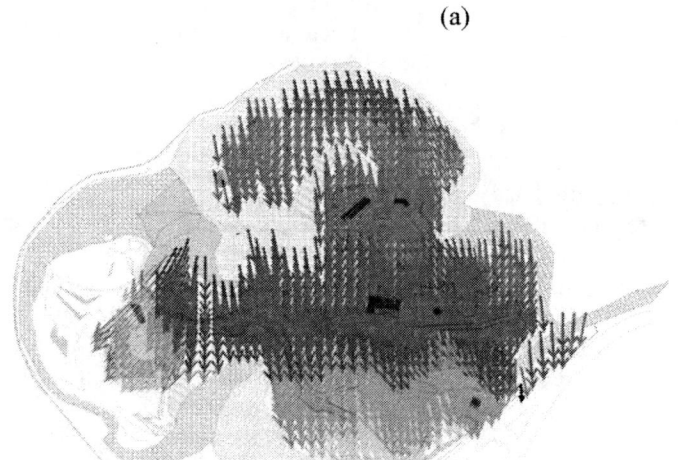

(b)

FIGURE 2. (a) Wind flow field – Area A
(b) Wind flow field – Area B

Taking into account the results of the numerical simulation for wind velocities, the PMV and PPD indicators are used in order to quantify the thermal comfort of the examined places' visitors. Ensuant to the above analysis a thermal comfort table is formed in order to investigate if and where, thermal comfort problems exist, for both the cases.

TABLE 2. Thermal comfort indicators for Area A. (Indicative part of the whole table)

Season	Calculation Parameters	Calculation Space	PMV	PPD(%)	Comments
Winter	Average wind temperature: 8.7 °C Average relative humidity: 74.4% Average wind velocity: 3.0 m/s Clothing factor: 2.2 Activity factor: 1.2 Average radiation temperature: 10 °C	Cinema	-1.2	35.2	Cold Area. It doesn't affect the area's visitors because the use of the cinema is only for the summer months. No measures are needed.
Winter	Average wind temperature: 8.7 °C Average relative humidity: 74.4% Average wind velocity: 2.9 m/s Clothing factor: 1.2 Activity factor: 2.4 Average radiation temperature: 10 °C	NW side of tennis court	-0.8	18.5	Acceptable area. At the NW side of the examined area the wind velocity is high, so the use of windbreak is suggested.
Summer	Average wind temperature: 27.4 °C Average relative humidity: 45.4% Average wind velocity: 3.0 m/s Clothing factor: 0.5 Activity factor: 1.2 Average radiation temperature: 24 °C	Cinema	-0.7	15.3	Acceptable area. No measures are needed.

After the PMV and PPD analysis, the architectural design solution is given based on the above numerical results. Dense planting, use of water sources, windbreaks, sun shadings and appropriate facing ground materials are used, based on bioclimatic principles.

CONCLUSIONS

This paper presents an attempt to bring CFD methods and numerical simulations to the attention of a wider group of architects and environmental scientists, for the evaluation of alternative open air space designs, using thermal comfort criteria. The work demonstrates that numerical solutions for such problems can be obtained quickly and economically, leading to a reliable estimation of PMV and PPD indicators.

REFERENCES

1. K. Papakonstantinou, "Design of spaces and installation with thermal comfort criteria and indoor air quality using CFD methods" Ph.D. Thesis, NTUA, 2000.
2. K. Papakonstantinou, C.Kiranoudis and N.C.Markatos, "Numerical Simulation of Air Pollution Dispersion and Thermal Comfort Analysis in the Myceane Hall of the National Archaeological Museum of Athens", 4th ECCOMAS Computational Fluid Dynamics Conference Proceedings, Athens, 1998.
3. N.C. Markatos, and K.A. Pericleous, "Laminar and turbulent natural convection in an enclosed cavity." in *Int. J. Heat Mass Transfer*, 27, No.5, pp. 755-772, 1984.

Quasi Monte Carlo Approach with Uniform Separation for Solving of the Rendering Equation

Anton A. Penzov*, Ivan T. Dimov*,† and Stanislava S. Stoilova**

*Department of Parallel Algorithms, Institute for Parallel Processing, Bulgarian Academy of Sciences, Acad. G. Bonchev Str., bl. 25 A, 1113 Sofia, Bulgaria
†ACET Centre, University of Reading, Whiteknights, PO Box 217, Reading, RG6 6AH, UK
**Department of Computational Mathematics, Institute of Mathematics and Informatics, Bulgarian Academy of Sciences, Acad. G. Bonchev Str., bl. 8, 1113 Sofia, Bulgaria

Abstract. This paper is directed to the advanced parallel Quasi Monte Carlo (QMC) methods for realistic image synthesis. We propose and consider a new QMC approach for solving the rendering equation with uniform separation. First, we apply the symmetry property for uniform separation of the hemispherical integration domain into 24 equal sub-domains of solid angles, subtended by orthogonal spherical triangles with fixed vertices and computable parameters. Uniform separation allows to apply parallel sampling scheme for numerical integration. Finally, we apply the stratified QMC integration method for solving the rendering equation. The superiority our QMC approach is proved.

Keywords: Quasi Monte Carlo, Uniform Separation, Stratified Sampling, Rendering Equation, Image Synthesis.

PACS: 02.70.Uu Applications of Monte Carlo methods; 02.60.Cb Numerical simulation, solution of equations; 02.30.Rz Integral equations.
MSC2000: 68U05 Computer graphics; 65C05 Monte Carlo methods; 65C20 Models, numerical methods; 68Q10 Models of computation (nondeterministic, parallel, interactive, probabilistic, etc.); 65R20 Integral equations; 45B05 Fredholm integral equations.

INTRODUCTION

The realistic image synthesis is the main task in the field of computer graphics. The photorealistic image creation requires the proper estimation of the global illumination in a virtual described scene. From mathematical point of view, it is equivalent to the solution of the rendering equation [1]. The rendering equation is a second kind Fredholm type integral equation, describing the light propagation in a scene (see in Fig. 1). The radiance L, leaving from a point x on the surface of the scene in direction $\omega \in \Omega_x$, where Ω_x is the hemisphere at point x, is the sum of the self radiating light source radiance L^e and all reflected radiance: $L(x,\omega) = L^e(x,\omega) + \int_{\Omega_x} L(h(x,\omega'), -\omega') f_r(-\omega', x, \omega) \cos \theta' d\omega'$,

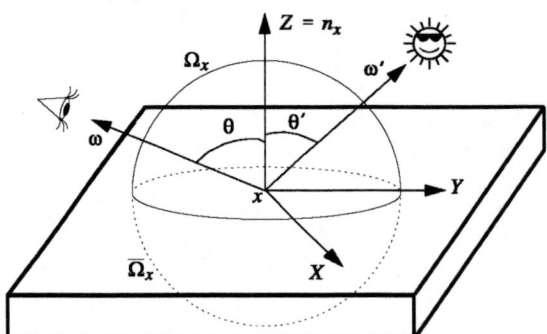

FIGURE 1. The geometry for the rendering equation

where $h(x,\omega')$ is the first point that is hit when shooting a ray from x into direction ω'. The radiance L^e has non-zero value if the considered point x is a point from solid light source. Therefore, the reflected radiance in direction ω is an integral of the radiance incoming from all points, which can be seen through the hemisphere Ω_x at point x attenuated by the surface BRDF (Bidirectional Reflectance Distribution Function) $f_r(-\omega', x, \omega)$ and the projection $\cos \theta'$. The angle θ' is the angle between the surface normal n_x at point x and the direction ω'. The law for energy conservation holds, i.e.: $\int_{\Omega_x} f_r(-\omega', x, \omega) \cos \theta' d\omega' < 1$, because a real scene always reflects less light than it receives from the light sources due to light absorption of the objects.

When the point x is a point from a transparent object the transmitted light component must be added to the rendering equation. This component estimates the total light transmitted trough the object and incoming to the point x from all directions opposite to the hemisphere Ω_x. The transmitted light in direction ω is an integral similar to the the reflected radiance integral where the domain of integration is the hemisphere $\overline{\Omega}_x$ at point x and BRDF is substituted by the surface BTDF (Bidirectional Transmittance Distribution Function) [2]. In this case the integration domain for solving the rendering equation is a sphere $\Omega^{(x)}$ at point x, where $\Omega^{(x)} = \Omega_x \bigcup \overline{\Omega}_x$.

The global illumination can be modeled as stationary linear iterative process [3], where multi-dimensional integrals are considered. It is very difficult or sometimes impossible the rendering equation to be solved analytically. Frequently, the Monte Carlo (MC) methods for numerical integration are the only practically effective method for multi-dimensional integrals.

UNIFORM SEPARATION OF INTEGRATION DOMAIN

Uniform separation (US) of integration domain for the parallel solution of the rendering equation is introduced by us in [4]. Sampling the unit hemisphere or sphere requires generation of sampling directions and it is enough to generate points over them. Consider hemisphere or sphere with center at the origin of a Descartes coordinate system. Similarly to the Bresenham algorithm for raster display of circle we apply the symmetry property for partitioning hemisphere and sphere. It is obvious that the coordinate planes partition the hemisphere into 4 equal areas and the sphere into 8 equal areas. The partitioning of each one area into sub-domains can be continued by the three bisector planes. One can see that the bisector planes to the dihedral angles $(\vec{X}, \vec{Y})$, $(\vec{X}, \vec{Z})$ and $(\vec{Z}, \vec{Y})$, partition each area into 6 equal sub-domains. In Fig. 2(a) we show the partitioning of the area with positive coordinate values of X, Y and Z into 6 equal sub-domains. In this way we can partition the hemisphere into 24 and the sphere into 48 equal sub-domains, respectively. Something more, due to the planes of partitioning, each sub-domain is symmetric to all others. The symmetric property allows us to calculate in parallel the coordinates of the symmetric points. To calculate the coordinates of point A_2 we consecutively multiply the coordinates of point $A_0(x_0, y_0, z_0)$ by two matrix of symmetry:

$$A_0(x_0, y_0, z_0) * \begin{pmatrix} 0 & 0 & 1 \\ 0 & 1 & 0 \\ 1 & 0 & 0 \end{pmatrix} * \begin{pmatrix} 1 & 0 & 0 \\ 0 & 0 & 1 \\ 0 & 1 & 0 \end{pmatrix} = A_1(z_0, y_0, x_0) * \begin{pmatrix} 1 & 0 & 0 \\ 0 & 0 & 1 \\ 0 & 1 & 0 \end{pmatrix} = A_2(z_0, x_0, y_0).$$

Therefore, sampling the hemisphere and once calculating the coordinates of a sampling point $A_0(x_0, y_0, z_0)$ from a given sub-domain, we can calculate in parallel the other sampling point coordinates for the hemisphere. The coordinate of symmetric points when we sampling the sphere can be calculated in a same way and only differ in negative sign of the Z coordinate. This kind of partitioning allows to sample only one sub-domain and to calculate in parallel all other samples for the hemisphere or sphere. Since the symmetry is identity, the generation of uniformly distributed random samples in a sub-domain leads to the uniform distribution of all samples in the hemisphere and sphere.

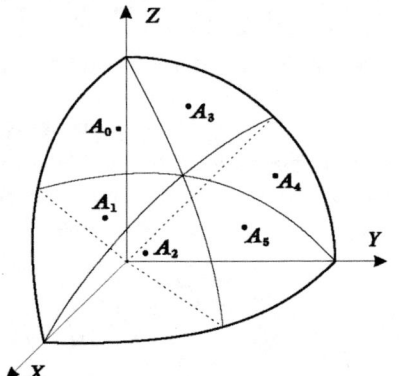
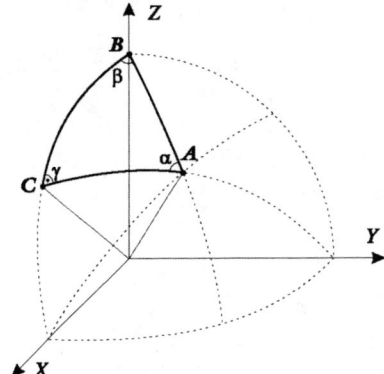

FIGURE 2. (a) Partition of a spherical area into 6 sub-domains   (b) Spherical triangle $\Omega_{\triangle ABC}$ geometry

Let us now apply the partitioning of the hemisphere Ω_x into 24 non-overlapping equal size sub-domains of orthogonal spherical triangles Ω_{i_x} (see in Fig. 2(b)), where $\Omega_{\triangle ABC} = \Omega_{i_x} = \frac{1}{24}\Omega_x = \frac{\pi}{12}$ for $i_x = 1, 2, \ldots, 24$. We can rewrite the rendering equation as: $L(x, \omega) = L^e(x, \omega) + \sum_{i_x=1}^{24} \int_{\Omega_{i_x}} L(h(x, \omega'), -\omega') f_r(-\omega', x, \omega) \cos \theta' d\omega'$, where $\Omega_x = \bigcup_{i_x=1}^{24} \Omega_{i_x}$. Therefore, the solution of rendering equation can be find as a sum of integrals over equal size non-overlapping sub-domains Ω_{i_x}. Consider the probability $p = \int_{\Omega_x} p(\omega') d\omega' = \sum_{i_x=1}^{24} \int_{\Omega_{i_x}} p(\omega') d\omega' = \sum_{i_x=1}^{24} p_{i_x} = 1$, it is obvious that $p_{i_x} = \int_{\Omega_{i_x}} p(\omega') d\omega' = \frac{1}{24}$ for $i_x = 1, 2, \ldots, 24$. Each sub-domain is sampled by random points $N_{i_x} \in \Omega_{i_x}$ with a density function $p(\omega')/p_{i_x}$. For all sub-domains N, independent sampling points are generated in parallel using the sampling scheme for hemisphere, where $N = 24 N_{i_x}$ for $i_x = 1, 2, \ldots, 24$, or the sampling points are equal numbers in each sub-domain.

For solving multi-dimensional integrals with uniform separation (see [5]), the total domain of integration $\Omega_x^{(s)}$ can be represented as $\Omega_x^{(s)} = \Omega_{x_1} \times \Omega_{x_2} \times \ldots \times \Omega_{x_s} = \prod_{j=1}^{s} \left(\bigcup_{i_{x_j}=1}^{24} \Omega_{i_{x_j}} \right) = 24^s \left(\frac{\pi}{12} \right)^s$. Let us suppose that $s = k_\varepsilon$ is maximum level of recursion (recursion depth or number of iterations) sufficient for numerical solution of the integral with a desired truncation error ε. In this case, on each iteration we have to solve the multi-dimensional integrals of type:

$$L^{(k_\varepsilon)} = L_{k_\varepsilon} - L_{k_\varepsilon-1} = \sum_{i_{x_1}=1}^{24} \cdots \sum_{i_{x_{k_\varepsilon}}=1}^{24} \int_{\Omega_{i_{x_1}}} \cdots \int_{\Omega_{i_{x_{k_\varepsilon}}}} L^e(x_{k_\varepsilon+1}, \omega'_{k_\varepsilon}) F(\omega'_1, \ldots, \omega'_{k_\varepsilon}) d\omega'_1 \ldots d\omega'_{k_\varepsilon},$$

where $F(\omega'_1, \ldots, \omega'_{k_\varepsilon}) = \prod_{j=1}^{k_\varepsilon} K_j(x_j, \omega'_j)$ and $K_j(x_j, \omega'_j) = f_r(-\omega'_j, x_j, \omega'_{j-1}) \cos\theta'_j$ for $j = 1, \ldots, k_\varepsilon$ as $L_0 = L^e(x_1, \omega)$ (note that $L_0 = L^e(x_1, \omega) = 0$ if the point x_1 is not a point from solid light source).

QUASI MONTE CARLO APPROACH WITH UNIFORM SEPARATION

Applying quasi-random sampling points for numerical integration of rendering equation we have QMC method. QMC method promises a faster convergence rate than the crude MC method due to the usage of deterministic sampling points. In computer graphics for QMC solution of the rendering equation [6], the most frequently applied ones are low discrepancy Halton sequences and finite Hammersley point sets as sampling points. The estimation of the integral approximation error ε is determined by the Koksma-Hlawka inequality:

$$\left| \int_{\Omega_x^s} g(x) dx - \frac{1}{N} \sum_{i=0}^{N-1} g(x_i) \right| \leq D_N^*(P_N) Var[g(x)],$$

where $Var[g(x)]$ is the variance of the kernel $g(x)$ of the rendering equation in the sense of Hardy-Krause and $D_N^*(P_N)$ is the discrepancy or the measure for the deviation of the point set $P_N = \{x_0, x_1, \ldots, x_{N-1}\}$ from uniform distribution. It is known that the discrepancy Halton sequences and finite Hammersley point sets are bounded by:

$$D_N^*(P_N^{Halton}) < \frac{s}{N} + \frac{1}{N} \prod_{j=1}^{s} \left(\frac{b_j-1}{2\log b_j} \log N + \frac{b_j+1}{2} \right) \quad \text{and} \quad D_N^*(P_N^{Hammersley}) < \frac{s}{N} + \frac{1}{N} \prod_{j=1}^{s-1} \left(\frac{b_j-1}{2\log b_j} \log N + \frac{b_j+1}{2} \right),$$

where s is the dimension of the points and $b_1, b_2, \ldots, b_s$ are the first s prime numbers, so as to reduce correlations between the components.

Let us apply **uniform separation** of the hemispherical domain $\Omega_x = \bigcup_{i_x=1}^{24} \Omega_{i_x}$ and use N' number of low discrepancy sampling points with a probability $p_{i_x} = \frac{1}{24}$ in each sub-domain Ω_{i_x}. Therefore, the total number of sampling points for the domain Ω_x is $N = 24N'$. The solution of rendering equation can be find as a sum of integrals over equal size non-overlapping sub-domains Ω_{i_x} and applying Koksma-Hlawka inequality we obtain:

$$\left| \int_{\Omega_x} g(x) dx - \frac{1}{N} \sum_{i=0}^{N-1} g(x_i) \right| = \left| \sum_{i_x=1}^{24} \left(\int_{\Omega_{i_x}} g(x) dx - \frac{p_{i_x}}{N'} \sum_{i=0}^{N'-1} g(x_i) \right) \right| \leq \sum_{i_x=1}^{24} \left| \int_{\Omega_{i_x}} g(x) dx - \frac{p_{i_x}}{N'} \sum_{i=0}^{N'-1} g(x_i) \right| \leq D_{N'}^*(P_{N'}) \sum_{i_x=1}^{24} Var\left[g^{(\Omega_{i_x})}(x) \right],$$

where $Var\left[g^{(\Omega_{i_x})}(x) \right]$ is the variance in the sub-domain Ω_{i_x} and $D_{N'}^*(P_{N'})$ is the discrepancy of the first N' points from the low discrepancy point set $P_N = \{x_0, x_1, \ldots, x_{N-1}\}$. The above result can be generalized for domains of arbitrary dimension $\Omega_x^{(s)} = \Omega_{x_1} \times \Omega_{x_2} \times \ldots \times \Omega_{x_{k_\varepsilon}}$ and $k_\varepsilon = s \geq 2$:

$$\left| \int_{\Omega_x^s} g(x)dx - \frac{1}{N} \sum_{i=0}^{N-1} g(x_i) \right| \leq D_{N'}^*(P_{N'}) \sum_{i_x=1}^{24^s} Var\left[g^{(\Omega_{i_x})}(x) \right].$$

According to [7] and to our previous results obtained in [5] for MC solution of the rendering equation with **uniform separation - US**, it can be shown that $\sum_{i_x=1}^{24^s} Var\left[g^{(\Omega_{i_x})}(x) \right] = Var\left[g^{US}(x) \right] \leq Var[g(x)]$ and $Var\left[g^{US}(x) \right] = \frac{1}{24} Var[g(x)]$ for one-dimensional case, respectively $Var\left[g^{US}(x) \right] = \frac{1}{24^s} Var[g(x)]$ for arbitrary dimension of $s \geq 1$. Substituting this result for the variance in the above inequality and taking into account the Neumann series, one can see that at QMC solution of the rendering equation with **uniform separation** we obtain better error estimation:

$$\sum_{j=1}^{s} \left| \int_{\Omega_x^s} g(x)dx - \frac{1}{N} \sum_{i=0}^{N-1} g(x_i) \right| \leq \left(\sum_{j=1}^{s} \frac{1}{24^s} D_{N'}^*(P_{N'}) \right) Var[g(x)].$$

In order to show the superiority of proposed QMC approach with **uniform separation** to classical QMC, the inequality $\left(\frac{1}{24^s} \right) D_{N'}^*(P_{N'}) < D_N^*(P_N)$ must be hold for any dimension $s \geq 1$. Consider the one-dimensional case with low discrepancy Halton sequence and let us suppose that *the difference* $\left(\frac{1}{24} \right) D_{N'}^*(P_{N'}^{Halton}) - D_N^*(P_N^{Halton}) \geq 0$ *is not negative*. Recall that $N = 24N'$ and applying the above mentioned inequality for bounding the discrepancy of Halton sequence, we can write the following:

$$0 \leq \frac{1}{24} D_{N'}^*\left(P_{N'}^{Halton} \right) - D_N^*\left(P_N^{Halton} \right) < \frac{1}{24N'} - \frac{1}{N} + \frac{1}{24N'} \left(\frac{b-1}{2\log b} \log N' + \frac{b+1}{2} \right) - \frac{1}{N} \left(\frac{b-1}{2\log b} \log N + \frac{b+1}{2} \right)$$

$$= \frac{1}{24N'} \left(\frac{b-1}{2\log b} \right) (\log N' - \log N) = \frac{1}{24N'} \left(\frac{b-1}{2\log b} \right) (\log N' - \log 24N') = -\frac{1}{24N'} \left(\frac{b-1}{2\log b} \right) \log 24 < 0,$$

which is contradiction to the above assumption for *strictly positive difference*, because b is a prime number and N' is also positive integer. Similar considerations could be taken for the multi-dimensional case too, as well as to the discrepancy of Hammersley point sets. Always the result is one and the same for arbitrary dimension. Therefore, the superiority of proposed QMC with **uniform separation** is proved.

CONCLUSION

A new QMC approach for solving the rendering equation with US is proposed. The estimate of the rate of convergence is obtained using the stratified QMC method. The uniform separation leads to convergence improvement of the QMC method as well as allows easy parallel realization. In our proposal the sampling points are reused in a natural way leading to reduction of the generated low discrepancy sampling points, which is another advantage of the method.

ACKNOWLEDGMENTS

This paper is supported by FP6 INCO Grand 016639/2005, Project - BIS-21++ and by the Ministry of Education and Science of Bulgaria under Grand No. I-1405/04.

REFERENCES

1. J.T. Kajiya, "The Rendering Equation", *ACM Computer Graphics (Proceedings of Siggraph '86)*, **20**(4), pp. 143–150, (1986).
2. Ph. Dutré, *Global Illumination Compendium*, Script 2003, http://www.cs.kuleuven.ac.be/ ~phil/GI/TotalCompendium.pdf
3. L. Szirmay-Kalos, *Monte-Carlo Methods in Global Illumination*, Script 1999/2000, http://www.fsz.bme.hu/ szirmay/script.pdf
4. I.T. Dimov, A.A. Penzov and S.S. Stoilova, "Parallel Monte Carlo Sampling Scheme for Sphere and Hemisphere", NMA 2006, T. Boyanov et al. (Eds.), LNCS **3401**, pp. 148–155, Springer-Verlag Berlin Heidelberg, (2007).
5. I.T. Dimov, A.A. Penzov and S.S. Stoilova, "Parallel Monte Carlo Approach for Integration of the Rendering Equation", NMA 2006, T. Boyanov et al. (Eds.), LNCS **3401**, pp. 140–147, Springer-Verlag Berlin Heidelberg, (2007).

6. A. Keller, "Quasi-Monte Carlo Methods in Computer Graphics: The Global Illumination Problem", *Lectures in Applied Mathematics*, **32**, pp. 455–469, (1996).
7. E. Veach, *Robust Monte Carlo Methods for Light Transport Simulation* Ph.D. Dissertation, Stanford University, Dec. (1997).

Unsteady Boundary Layer Flow and Heat Transfer Over a Stretching Sheet

Cornelia Revnic[*], Teodor Grosan[†] and Ioan Pop[1†]

[*] "Tiberiu Popoviciu" Institute of Numerical Analysis, P.O.Box. 68-1, 400110, Cluj-Napoca, Romania
[†] Babes-Bolyai University, Applied Mathematics, R-3400 Cluj, CP 253, Romania

Abstract. Unsteady two-dimensional boundary layer flow and heat transfer over a stretching flat plate in a viscous and incompressible fluid of uniform ambient temperature is investigated in this paper. It is assumed that the plate is isothermal and is stretched in its own plane. Using appropriate similarity variables, the basic partial differential equations are transformed into a set of two ordinary differential equations. These equations are solved numerically for some values of the governing parameters, using Rungge-Kutta method of fourth order. Flow and heat transfer characteristics are determined and represented in some tables and figures. It is found that the structure of the boundary layer depends on the ratio of the velocity of the potential flow near the stagnation point to that of the velocity of the stretching surface. In addition, it is shown that the heat transfer from the plate increases when the Prandtl number increases. Our results are shown to include the steady situation as a special case considered by other authors. Comparison with known results is very good.

Keywords: heat transfer, stretching surface, the external inviscid flow, stagnation-point flow, boundary layer.
PACS: 02.60Lj, 44.20+b, 44.27+g, 47,10Ad, 47.85-g

INTRODUCTION

The unsteady boundary layers are important in several physical problems in aero - nautics, missile dynamics, acoustics etc. The work in this area was initiated by Moore [1], Lighthill [2] and Lin [3]. Critical reviews of unsteady boundary layers were presented by Stuart [4], Riley [5], Telionis [6], [7] and Pop [8]. In recent years certain aspects of the unsteady flows were investigated by Ma and Hui [9] and Ludlow et al. [10] using the classical method of Lie-group. The essence of the Lie-group method is that each of the variables in the initial equation is subjected to an infinitesimal transformation and the demand that the equation is invariant under these transformations leads to the determination of the possible symmetries (see Ludlow et al. [10]). The fundamental governing equations for fluid mechanics are the Navier-Stokes equations. This nonlinear set of partial differential equations have no general solutions, and only a small number of exact solutions have been found (see Wang [11]). Exact solutions are important for the following reasons: (i) the solutions represent fundamental fluid-dynamic flows. Also, owing to the uniform validity of exact solutions, the basic phenomena described by the Navier-Stokes equations can be more closely studied. (ii) the exact solutions serve as standards for checking the accuracies of the many approximate methods, whether they are numerical, asymptotic, or empirical.

Flow of a viscous fluid over a stretching sheet has an important bearing on several technological processes. In particular in the extrusion of a polymer in a melt-spinning process, the extruded from the die is generally drawn and simultaneously stretched into a sheet which is then solidified through quenching or gradual cooling by direct contact with water. Further, glass blowing, continuous casting of metals and spinning of fibres involve the flow due to a stretching surface, see Lakshmisha et al. [12]. In all these cases, a study of the flow field and heat transfer can be of significant importance since the quality of the final product depends to a large extent on the skin friction coefficient and the surface heat transfer rate. Crane [13] presented a simple closed form exponential solution of the steady two-dimensional flow caused solely by a linearly stretching sheet in an otherwise quiescent incompressible fluid. The simplicity of the geometry and the possibility of obtaining further exact solutions through simple generalizations have generated a lot of interest in extending it to more general situations. Such extensions include consideration of more general stretching velocity, application to non-Newtonian fluids, and inclusion of other physical effects such as suction or blowing, magnetic fields, etc. Unsteady two-dimensional boundary layer flow over a stretching surface

---

[1] Corresponding author: Tel.:-40-264594315; fax:+40-264591906; E-mail adress: pop.ioan@yahoo.co.uk (Pop Ioan)

CP1046, *Selected Papers from ICNAAM 2007 and ICCMSE 2007*
edited by T. E. Simos, G. Maroulis, G. Psihoyios, and Ch. Tsitouras
© 2008 American Institute of Physics 978-0-7354-0574-5/08/$23.00

has been studied by Na and Pop [14], Wang et al. [15], Elbashbeshy and Badiz [16], Sharidan et al. [17] and Ali and Magyari [18], while Lakshmisha et al. [12], Devi et al. [19] and Takhar et al. [20] have considered the unsteady three-dimensional-flow due to the impulsive motion of a stretching surface. The aim of the present analysis is to study the unsteady flow and heat transfer in the stagnation-point flow on a heated stretched surface in a viscous and incompressible fluid when both velocities of the stretching sheet and of the external flow (inviscid flow) are proportional to the distance from the stagnation-point and inversely to time. The geometry is similar to that proposed by Mahapatra and Gupta [21] for the steady two-dimensional stagnation-point flow towards a stretching sheet. The parabolic partial differential equations governing the flow and heat transfer have been reduced to a system of two ordinary differential equations which are solved using an implicit finite-difference scheme in combination with the shooting method.

PROBLEM FORMULATION

We consider the unsteady two-dimensional forced convection flow and heat transfer of a viscous and incompressible fluid near a stagnation point on a surface coinciding with the plain $y = 0$, the flow being confined to $y > 0$. Two equal and opposite forces are applied along the x - axis at the initial time $t = 0$, so that the surface is stretched keeping the origin fixed as shown in Fig.1. It is assumed that the uniform temperature of the plane is T_w, while the temperature of the ambient fluid is T_∞, where $T_w > T_\infty$ (heated plate). It is also assumed that the viscous dissipation effects are neglected. Under these assumptions, the system of boundary layer equations are given by

$$\frac{\partial u}{\partial x} + \frac{\partial v}{\partial y} = 0 \tag{1}$$

$$\frac{\partial u}{\partial t} + u\frac{\partial u}{\partial x} + v\frac{\partial u}{\partial y} = \frac{\partial u_e}{\partial t} + u_e\frac{\partial u_e}{\partial x} + v\frac{\partial^2 u}{\partial y^2} \tag{2}$$

$$\frac{\partial T}{\partial t} + u\frac{\partial T}{\partial x} + v\frac{\partial T}{\partial y} = \alpha\frac{\partial^2 T}{\partial y^2} \tag{3}$$

subject to the initial and boundary conditions are of the form:

$$
\begin{aligned}
t \quad &< 0: u = 0, v = 0, T = T_\infty \ for \ any \ y > 0 \\
t \quad &> 0: u = u_w(t,x), v = 0, T = T_w \ for \ y = 0 \\
t \quad &= 0: u = u_{ws}(x), v = 0, T = T_w \\
& \quad u \to u_e(t,x), T \to T_\infty \ as \ y \to \infty
\end{aligned} \tag{4}
$$

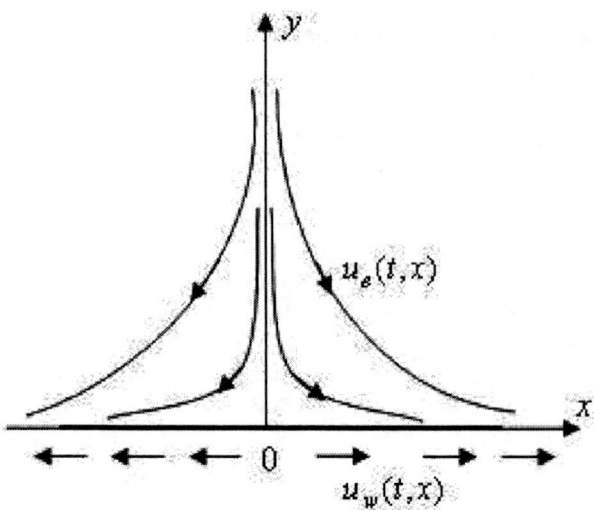

FIGURE 1. Physical model and coordinate system.

where u and v are the velocity components along the $x-$ and $y-$ axis, T is the fluid temperature, v is the kinematic viscosity, $u_{ws} = cx$ (c is a positive constant) and α is the thermal diffusivity. Following Surma Devi [19] et al., we assumed that $u_w(t,x)$ and $u_e(t,x)$ are given by

$$u_w(t,x) = \frac{cx}{(1-\gamma t)}, \quad u_e(t,x) = \frac{ax}{(1-\gamma t)} \tag{5}$$

where a is a positive constant. The momentum and energy equations can be transformed to the corresponding ordinary differential equations by the following substitutions:

$$\begin{aligned}
\psi &= (cv/(1-\gamma t))^{1/2}xf(\eta), \\
\theta(\eta) &= (T - T_\infty)/(T_w - T_\infty), \\
\eta &= (c/v(1-\gamma t))^{1/2}y
\end{aligned} \tag{6}$$

where ψ is the stream function which is defined in the usual way as $u = \partial\psi/\partial y$ and $v = -\partial\psi/\partial x$. Substituting (6) into Eqs. (2) and (3), we obtain the following two ordinary differential equations:

$$f''' + f f'' + \frac{a^2}{c^2} - f'^2 + \frac{\gamma}{c}\left(\frac{a}{\epsilon} - \frac{\eta}{2}f'' - f'\right) = 0 \tag{7}$$

$$\frac{1}{Pr}\theta'' + f\theta' - \frac{\gamma}{2c}\eta\theta' = 0 \tag{8}$$

subject to the boundary conditions (4) which become

$$f(0) = 0, f'(0) = 1, \theta(0) = 1 \tag{9}$$

$$f'(\infty) = \frac{a}{c}, \theta(\infty) = 0 \tag{10}$$

where Pr is the Prandtl number and primes denote differentiation with respect to η.
The physical quantities of interest are the skin friction coefficient C_f and the local Nusselt number Nu_x, which are defined as

$$C_f = \frac{\tau_w}{\rho u_{ws}^2}, \quad Nu_x = \frac{xq_w}{k(T_w - T_\infty)}, \tag{11}$$

where τ_w is the skin friction and q_w is the heat transfer from the plate which are given by

$$\tau_w = \mu\left(\frac{\partial u}{\partial y}\right)_{y=0}, \quad q_w = -k\left(\frac{\partial T}{\partial y}\right)_{y=0} \tag{12}$$

with μ and k being the dynamic viscosity and thermal conductivity, respectively. Using (6), we get

$$(1-\gamma t)^{3/2}Re_x^{1/2}C_f = f''(0), \tag{13}$$
$$(1-\gamma t)^{1/2}Re_x^{-1/2}Nu_x = -\theta'(0)$$

Where $Re = (cx)x/v$ is the low Reynolds number. It is important to notice that for the steady-state case, Eqs. (7) and (8) reduced to

$$f''' + f f'' - f'^2 + \frac{a^2}{c^2} = 0 \tag{14}$$

$$\frac{1}{Pr}\theta'' + f\theta' = 0 \tag{15}$$

with the boundary conditions (9)-(10). Equations (14) and (15) with the boundary conditions (9)-(10) where established by Mahapatra and Gupta [21].

TABLE 1. Values of $f''(0)$ for some values of a/c when the flow is steady. ( ) values reported by Mahapatra and Gupta [21].

a/c	0.10	0.20	0.50	2.00
$f''(0)$	-0.9696	-0.9182	-0.6673	2.0175
	(-0.9694)	(-0.9181)	(-0.6673)	(2.0175)

TABLE 2. Values of $\theta'(0)$ for some values of a/c and Pr when the flow is steady. ( ) values reported by Mahapatra and Gupta [21].

a/c / Pr	0.05	0.5	1	1.5
0.1	-0.081	-0.381	-0.603	-0.777
	(-0.081)	(-0.383)	(-0.603)	(-0.777)
0.5	-0.137	-0.472	-0.691	-0.863
	(-0.136)	(-0.473)	(-0.692)	(-0.863)
2	-0.248	-0.711	-0.978	-1.171
	(-0.241)	(-0.709)	(-0.974)	(-1.171)

SOLUTION

The systems of ordinary differential equations (7)-(8) and (14)-(15) subject to the boundary condition (9)-(10) have been solved numerically for some values of the parameters a/c, t and Pr using Rungge-Kutta method of fourth order combined with the shooting technique. For the physical considearation we take $\gamma = -1$. Some values of $f''(0)$ and $\theta'(0)$ are given in Tables 1 and 2 for the case of the steady flow.

We can see from these tables that there is a very good agreement between our results and those obtained by Mahapatra and Gupta [21]. Therefor, we are confident that the results obtained using the present method are accurate.

Figures 2 - 5 show the velocities profiles f and f' along with the corresponding streamlines patterns for the case of unsteady flow, Eqs. (7) and (8). The values of the parameters are $a = 0.1$, $c = 1$ and $t = 0, 1, 2, 3$. It is interesting to notice that the solution of Eq. (7) is not unique. Thus, there are two solutions, one Fig. 2 representing an attached flow and the other one Fig. 4 the reversed flow. These is in agreement with the results obtained by Ma and Hui [9] for the unsteady two-dimensional boundary layer flow near a stagnation point of a fixed plat plate.

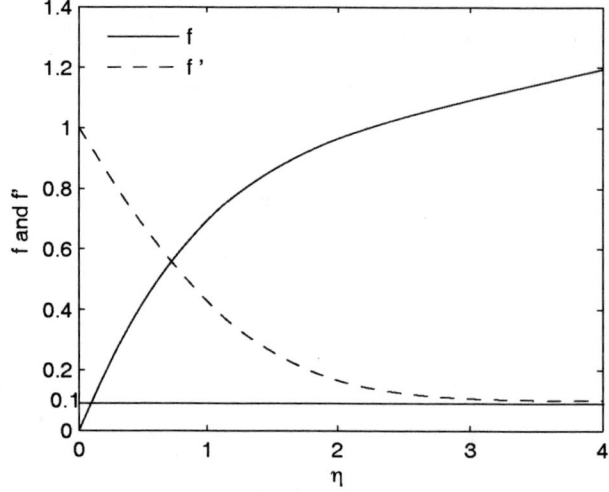

FIGURE 2. The first solution of $f(\eta)$ and $f'(\eta)$ for $a/c = 0.1$.

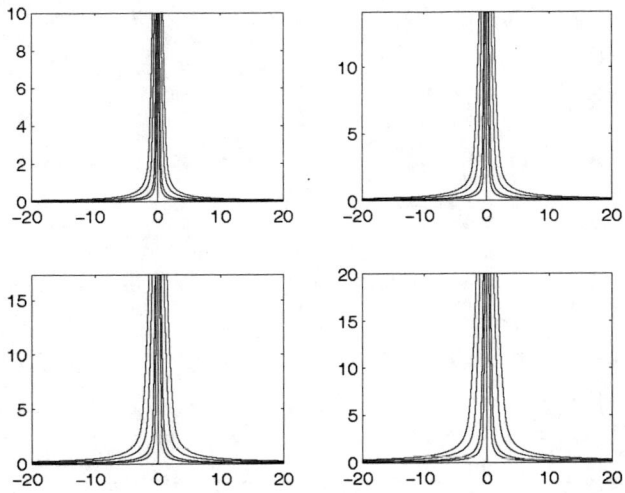

FIGURE 3. The streamlines function for: $t = 0, 1, 2$ and 3 corresponding to the first solution for $a/c = 0.1$.

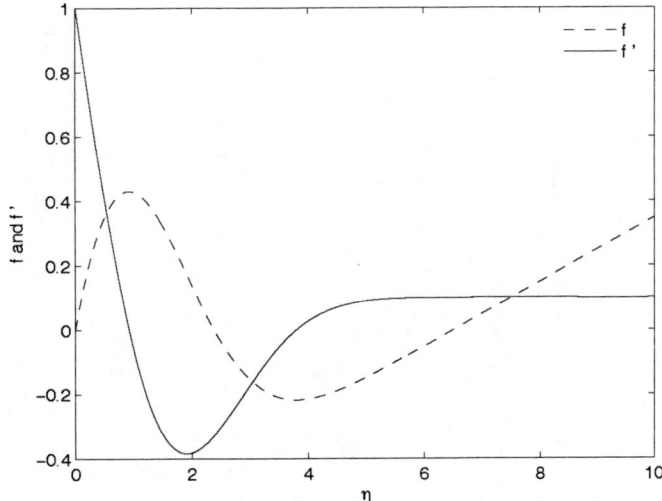

FIGURE 4. The second solution of $f(\eta)$ and $f'(\eta)$ for $a/c = 0.1$.

Fig. 6 illustrates the dimensionless temperature profiles $\theta(\eta)$ for some values of Pr when $a/c = 2$. We notice that temperature profile increase when Pr decreases. Further, Fig. 7 shows the variation of the heat transfer from the wall $-\theta'(0)$ with a/c and different values of Pr. It is evident from Fig. 7 that an increase in Pr result in a decrease in the thermal boundary layer thickness and as a consequence the heat transfer from the wall increases with Pr.

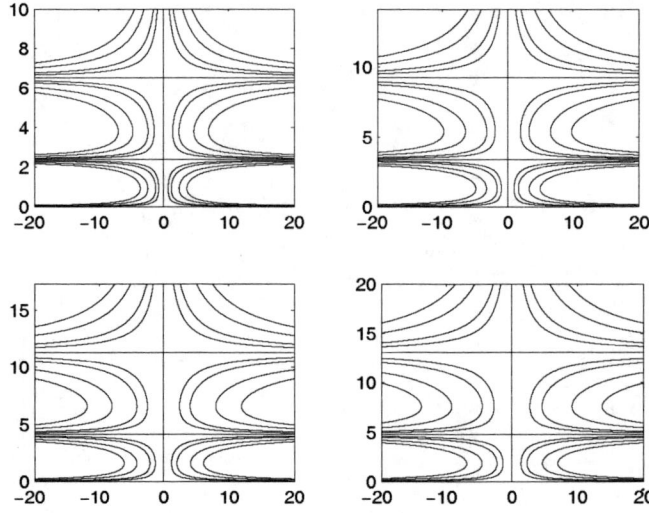

FIGURE 5. The streamlines function for: $t = 0, 1, 2$ and 3 corresponding to the first solution for $a/c = 0.1$.

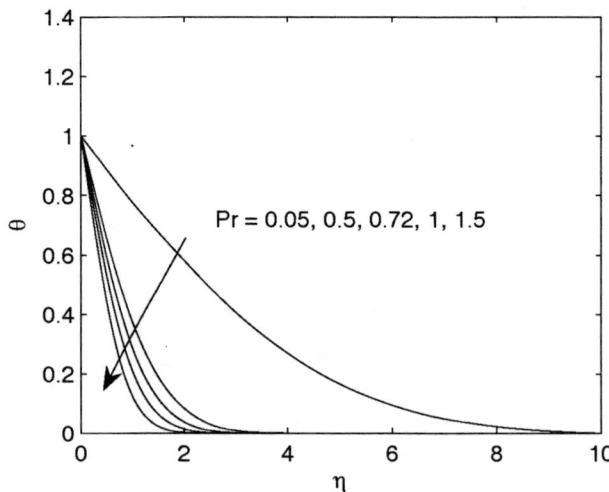

FIGURE 6. Temperature profiles of $\theta(\eta)$ for several values of Pr and $a/c = 2$ in respect with η.

CONCLUSION

The unsteady two-dimensional stagnation-point flow and heat transfer of a viscous and incompressible fluid over an isothermal stretching flat plate in its own plane has been numerically analyzed in detailed. Following Surma Devi et al. [19] similarity variables where used to reduced the governing partial differential equations to ordinary differential equations. Solving numerically these equations, we have been able to determine the velocity and temperature profiles, skin friction and heat transfer from the plate. For the case of steady-state flow, we have compared our present results with those of Mahapatra and Gupta [21]. The agreement between the results is excelent. Effects of a/c and Pr on the flow and heat transfer characteristic have been examened and discussed in detail. It is shown that for small values of a/c the solution of the ordinary differential equation is not unique. One solution represents an attached flow and the other one a reversed flow. It shoud be noticed that we have determined solutions of the problem for more values of the governing parameters but in order to save space, the reported results are limited only to some values of these parameters.

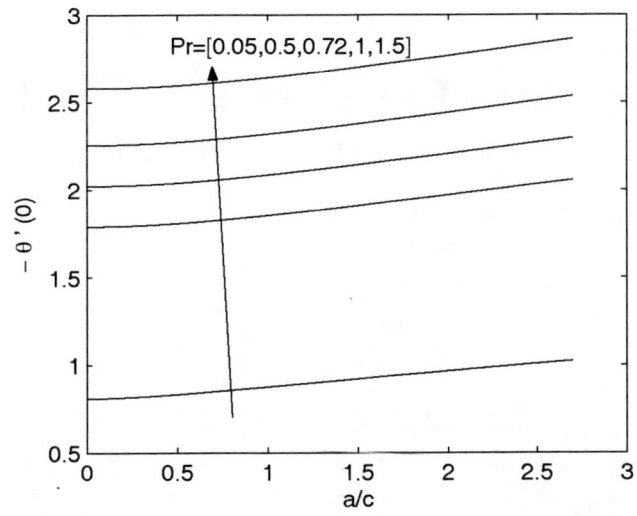

FIGURE 7. Variation of the heat transfer with a/c for several values of Pr.

REFERENCES

1. F. K. Moore, *Unsteady laminar boundary layer flow*, NACA TN, 1951, 3471.
2. M. J. Lightill, Proc. Roy. Soc. **224A**, 1–23 (1954).
3. C. C. Lin, Proc. 9th Int. Cong. Appl. Mech **4**, 155–168 (1956).
4. J. T. Stuart **1**, 1–40 (1971).
5. N. Riley, SIAM Review **17**, 274–297 (1975).
6. D. P. Telionis, J. Fluids Engng. **101**, 29–43 (1979).
7. D.P. Telionis, *Unsteady Viscous Flows,* Springer, Berlin, 1981.
8. I. Pop, *Theory of Unsteady Boundary Layers* (in Romanian), Ed. Stiintifica si Enciclopedica, Bucuresti–Romania, 1983.
9. P.K.H. Ma, and W.H. Hui, J. Fluid Mech **216**, 537–559 (1990).
10. D. K. Ludlow, P. A. Clarkson, and A.–P. Bassom, Q. Jl. Mech. Appl. Math. **53**, 175–206 (2000).
11. C. Y. Wang, Annu. Rev. Fluid Mech **23**, 159–177 (1991).
12. K. N. Lakshmisha, S. Venkateswaran, and G. Nath, ASME J. Heat Transfer **110**, 590–595 (1988).
13. L.J. Crane, J. Appl. Math. Phys. (ZAMP) **21**, 645–647 (1970).
14. I. Pop, and T.Y. Na, **23**, 413–422 (1996).
15. C. Y. Wang, Q. Du, M. Miklavčič, and C.–C. Chang, SIAM J. Appl. Math. **57**, 1–14 (1997).
16. E. M. A. Elbashbeshy, and M.A.A. Bazid, Heat Mass Transfer **41**, 1–4 (2004).
17. S. Sharidan, T. Mahmood, and I. Pop, Int. J. Appl. Mech. Engng., **11**, 647–654 (2006).
18. M. E. Ali, and E. Magyari, Int. J. Heat Mass Transfer **50**, 188–195 (2007).
19. C. D. Surma Devi, H.S. Takhar, and G. Nath, Int. J. Heat Mass Transfer **29**, 1996– 1999 (1986).
20. H. S. Takhar, A.J. Chamkha, and G. Nath, Acta Mechanica **146**, 59–71 (2001).
21. T. Ray Mahapatra, and A. S. Gupta, Heat Mass Transfer **38**, 517–521 (2002).

Parameter Depending Matuszewska-Orlicz Type Indices and their applications

N. Samko[1]

INTRODUCTION

Last decades there was observed an increasing interest to the study of function spaces whose characteristics may vary from point to point, measure metric spaces where the dimension depends on a point, being one of the examples. For such spaces, the Muckenhoupt conditions for radial weights, includes Zygmund type integral constructions of $\mu B(x, r)$. This led us to the study of Zygmund-Bary-Stechkin functions $w(x, r)$ depending on a parameter x belonging to an arbitrary metric measure space.

THE GENERALIZED BARY-STECHKIN CLASS Φ_γ^β

Let X be a metric space with a positive measure μ and Ω a bounded open set in X with $\ell = diam\,\Omega$, $0 < \ell < \infty$. We will deal with functions $w(x, r)$ defined on $\Omega \times [0, \ell]$ which are almost increasing (**a.i.**) in variable r, and we will be interested in properties related to this almost monotonicity, uniform with respect to x.

By $W = W(\Omega \times [0, \ell])$ we denote the class of functions w with the properties: $w \in L^\infty(\Omega \times [0, \ell])$; $w(x, r)$ is continuous in $r \in [0, \ell]$ for any fixed $x \in \Omega$; $w(x, 0) = 0$, but ess $\inf_{x \in \Omega} w(x, r) := d_0(r) > 0$ for every $r > 0$; for any fixed $x \in X$ the function $w(x, r)$ is a.i. in r uniformly in x. We also introduce the class
$$\widetilde{W}(X \times [0, \ell]) = \{w(x, r): \ \exists a = a(w) \in \mathbb{R}^1 \text{ such that } t^a w(x, t) \in W(X \times [0, \ell])\}.$$

Let $w \in W$. The numbers

$$m(w; x) = \sup_{r > 1} \frac{\ln\left(\liminf\limits_{h \to 0} \frac{w(x, rh)}{w(x, h)}\right)}{\ln r}, \quad M(w; x) = \inf_{r > 1} \frac{\ln\left(\limsup\limits_{h \to 0} \frac{w(x, rh)}{w(x, h)}\right)}{\ln r} \tag{1}$$

will be referred to as the *lower and upper index numbers* of a function $w \in W$ with respect to r. We refer to [2] for the indices $m(w), M(w)$ in the case where $w = w(r)$, compare these indices with the Matuszewska-Orlicz indices, see [3], p. 20.

Definition 1 *Let $\beta, \gamma \in \mathbb{R}^1$. By $\mathscr{Z}^\beta = \mathscr{Z}^\beta(\Omega \times [0, \ell])$, we denote the class of functions $w(x, r) \in W$ satisfying the condition*

$$\int_0^r \frac{w(x, t)}{t^{1+\beta}} dx \le A \frac{w(x, r)}{r^\beta}, \quad 0 < r \le \ell \tag{$\mathbb{Z}^\beta$}$$

and by $\mathscr{Z}_\gamma$ the class of functions $w(x, r) \in W$ satisfying the condition

$$\int_r^\ell \frac{w(x, t)}{t^{1+\gamma}} dt \le A \frac{w(x, r)}{r^\gamma}, \quad 0 < r \le \ell \tag{$\mathbb{Z}_\gamma$}$$

where $A = A(w) > 0$ does not depend on r and $x \in \Omega$. We denote $\Phi_\gamma^\beta = \mathbb{Z}^\beta \cap \mathbb{Z}_\gamma$.

For $w \in W(\Omega \times [0, \ell])$ we deal with the following known conditions:

---

[1] nsamko@ualg.pt

CP1046, *Selected Papers from ICNAAM 2007 and ICCMSE 2007*
edited by T. E. Simos, G. Maroulis, G. Psihoyios, and Ch. Tsitouras
© 2008 American Institute of Physics 978-0-7354-0574-5/08/$23.00

1) *Bary type conditions:*

$$\sum_{k=n+1}^{\infty} k^{\beta-1} w\left(x, \frac{1}{k}\right) \le A n^{\beta} w\left(x, \frac{1}{n}\right), \qquad (\mathbb{B}^{\beta})$$

2) *Lozinski type conditions:*

$$\exists\, C > 1 \quad \text{not depending on } x \text{ such that} \quad \underset{x \in \Omega}{\operatorname{ess\,inf}} \lim_{r \to 0} \frac{w(x, Cr)}{C^{\beta} w(x, r)} > 1, \qquad (\mathbb{L}^{\beta})$$

3) *Stechkin type conditions:*

$$\text{there exists a } \delta > 0 \text{ such that } \frac{w(x, r)}{r^{\beta+\delta}} \text{ is uniformly a.i.} \qquad (\mathbb{S}^{\beta})$$

Theorem 1 *Let $w \in \widetilde{W}(\Omega \times [0, \ell])$. Then conditions $(\mathbb{B}^{\beta}), (\mathbb{L}^{\beta}), (\mathbb{Z}^{\beta}), (\mathbb{S}^{\beta})$ are equivalent to each other and to the inequality $m(w) > \beta$, and condition $(\mathbb{S}^{\beta})$ holds with every $\delta < m(w) - \beta$.*

Theorem 2 *Let $w \in W$ and $0 \le \beta < \gamma < \infty$. Then*

$$w \in \Phi_{\gamma}^{\beta}(\Omega \times [0, \ell]) \iff \beta < m(w) \le M(w) < \gamma \qquad (2)$$

and for $w \in \Phi_{\gamma}^{\beta}$ and any $\varepsilon > 0$ there exist constants $c_1 = c_1(\varepsilon) > 0$ and $c_2 = c_2(\varepsilon) > 0$ not depending on $x \in \Omega$ such that

$$c_1 r^{M(w)+\varepsilon} \le w(x, r) \le c_2 r^{m(w)-\varepsilon}, \qquad 0 \le r \le \ell. \qquad (3)$$

APPLICATIONS

The above results may be used in the study of the mapping properties of singular and potential operators in generalized Hölder spaces $H^{\omega(x,\cdot)}(X)$, with variable characteristic, where X may be a metric measure space, defined by the norm

$$\|f\|_{H^{\omega(x,\cdot)}(X)} = \|f\|_{C(X)} + \sup_{x \in X} \sup_{0 < h < \ell} \frac{\omega(f, x, h)}{\omega(x, h)}, \quad \ell = \operatorname{diam} X,$$

where $\omega(f, x, h)$ is the local continuity modulus of a function f at the point x, and $\omega(x, h)$ stands for the dominant of these moduli.

We also consider an application of the index numbers depending on parameter to the problem of measuring local dimensions in metric measure spaces.

Local variable dimensions

Let (X, d, μ) be a metric measure space with quasidistance $d : X \times X \to \mathbb{R}^1$ and a positive measure μ with the condition $\operatorname{ess\,inf}_{x \in X} \mu B(x, r) > 0$ for every $r > 0$. The lower and upper local dimensions

$$\underline{dim}_X(x) = \liminf_{r \to 0} \frac{\ln \mu B(x, r)}{\ln r}, \quad \overline{dim}_X(x) = \limsup_{r \to 0} \frac{\ln \mu B(x, r)}{\ln r} \qquad (1)$$

may be attributed to a point x ([1]).

We introduce a new notion of *the local lower and upper dimensions* in the form

$$\underline{\eth im}_X(x) := \sup_{r > 1} \frac{\ln\left(\liminf\limits_{t \to 0} \frac{\mu B(x, rt)}{\mu B(x, t)}\right)}{\ln r}, \quad \overline{\eth im}_X(x) := \inf_{r > 1} \frac{\ln\left(\limsup\limits_{t \to 0} \frac{\mu B(x, rt)}{\mu B(x, t)}\right)}{\ln r}, \qquad (2)$$

An advantage of the usage of these local dimensions in comparison with dimensions (1) is in the fact that just in terms related to the dimensions $\underline{\eth im}_X(x)$ and $\overline{\eth im}_X(x)$ there may be given sufficient conditions for the validity of some integral inequalities involving the measures $\mu B(x,r)$, which appear when one considers the Muckenhoupt type A_p-condition on metric measure spaces.

We introduce the following version for the lower dimensions of the space:

$$\underline{\dim}(X) := \sup_{r>1} \frac{\ln\left(\liminf_{t\to 0}\operatorname{ess\,inf}_{x\in X}\frac{\mu B(x,rt)}{\mu B(x,t)}\right)}{\ln r} = \lim_{r\to 0}\frac{\ln\left(\limsup_{t\to 0}\operatorname{ess\,sup}_{x\in X}\frac{\mu B(x,rt)}{\mu B(x,t)}\right)}{\ln r}.$$

Similarly the upper ones are introduced.

Theorem 1 *Let* $u,v \in \widetilde{W}([0,\ell])$. *The conditions*

$$m(u) > -\underline{\dim}(X) \quad and \quad M(v) < \overline{\dim}(X) \tag{3}$$

are sufficient for inequalities

$$\int_0^h \frac{\mu B(x,r)u(r)}{r}\,dr \leq Cu(h)\mu B(x,h), \quad \int_0^h \frac{\mu B(x,r)}{rv(r)}\,dr \leq C\frac{\mu B(x,h)}{v(h)}.$$

Mapping properties of spherical potentials and hypersingular operators

Let $\mathbb{S}^{n-1}$ be the unit sphere in $\mathbb{R}^n$. We study spherical potential operators

$$(K^{\alpha(\cdot)}f)(x) = \int_{\mathbb{S}^{n-1}} \frac{f(\sigma)\,d\sigma}{|x-\sigma|^{n-1-\alpha(x)}}, \quad x\in\mathbb{S}^{n-1}, \tag{4}$$

where $0 < \Re\alpha(x) < 1$, and spherical hypersingular operators

$$(D^{\alpha(\cdot)}f)(x) = \lim_{\varepsilon\to 0}\int_{\substack{\mathbb{S}^{n-1}\\|x-\sigma|\geq\varepsilon}} \frac{f(\sigma)-f(x)}{|x-\sigma|^{n-1+\alpha(x)}}\,d\sigma, \quad x\in\mathbb{S}^{n-1}, \tag{5}$$

where $0 \leq \Re\alpha(x) < 1$. We consider these operators in the frameworks of the generalized Hölder spaces $H_0^{\omega(\cdot)}(\mathbb{S}_{n-1})$ with $\omega = \omega(x,h), 0 < h < 2$ depending on points x of the sphere. In particular, under the choice $\omega(x,h) = h^{\lambda(x)}$ we obtain the variable Hölder space which is usually denoted as $H^{\lambda(\cdot)}$. We admit functions $\omega(x,t)$ which belong to the Zygmund-Bary-Stechkin class in h uniformly in $x\in\mathbb{S}^{n-1}$.

Theorem 2 *Let* $\min_{x\in\mathbb{S}^{n-1}}\Re\alpha(x) > 0$, $\max_{x\in\mathbb{S}^{n-1}}\Re\alpha(x) < 1$ *and* $\alpha(x) \in H^{\omega_\alpha(\cdot)}(\mathbb{S}^{n-1})$. *Then the operator* $K^{\alpha(\cdot)}$ *is bounded from the space* $H^{\omega(\cdot)}(\mathbb{S}^{n-1})$ *with* $\omega(x,t) \in \Phi_{1-\Re\alpha(x)}$ *into the space* $H^{\omega_\alpha(\cdot)}(\mathbb{S}^{n-1})$.

Theorem 3 *Under conditions of Theorem 2, on* $\alpha(x)$ *the operator* $K^{\alpha(\cdot)}$ *is bounded from the space* $H^{\omega(\cdot)}(\mathbb{S}^{n-1})$ *into the space* $H^{\omega_\alpha(\cdot)}(\mathbb{S}^{n-1})$, *if* $\omega \in W(\mathbb{S}^{n-1}\times[0,2])$ *and the upper index number of the function* $\omega(x,\cdot)$ *satisfies the condition* $\operatorname{ess\,sup}_{x\in\mathbb{S}^{n-1}}\{M(\omega,x)+\Re\alpha(x)\} < 1$.

Theorem 4 *Let* $\min_{x\in\mathbb{S}^{n-1}}\Re\alpha(x) > 0$, $\max_{x\in\mathbb{S}^{n-1}}\Re\alpha(x) < 1$, $\alpha(x) \in H^{\omega-\alpha(\cdot)}(\mathbb{S}^{n-1})$. *and* $w(x,\cdot)$ *satisfies the condition* $t^{-\alpha(x)}w(x,t) \in \Phi^0$. *Then the operator* D^α *is bounded from the space* $H^{\omega(\cdot)}(\mathbb{S}^{n-1})$ *into the space* $H^{\omega-\alpha(\cdot)}(\mathbb{S}^{n-1})$.

A reformulation of Theorem 4 in terms of the lower index runs as follows.

Theorem 5 *Under conditions of Theorem 4, on* $\alpha(x)$, *the operator* D^α *is bounded from* $H^{\omega(\cdot)}(\mathbb{S}^{n-1})$ *to* $H^{\omega-\alpha}(\mathbb{S}^{n-1})$, *if* $t^{\Re\alpha(x)}w(x,t) \in W(\mathbb{S}^{n-1}\times[0,2])$ *and the lower index number of the function* $\omega(x,\cdot)$ *satisfies the condition* $\operatorname{ess\,inf}_{x\in\mathbb{S}^{n-1}}\{m(\omega,x)-\Re\alpha(x)\} > 0$.

ACKNOWLEDGMENTS

This work was supported by Research Grant SFRH/BPD/34258/2006, FCT, Portugal, and the INTAS project "Variable Exponent Analysis", Nr.06-1000017-8792.

REFERENCES

1. K. Falconer. *Techniques in fractal geometry*. John Wiley & Sons Ltd., Chichester, 1997.
2. N.K. Karapetiants and N.G. Samko. Weighted theorems on fractional integrals in the generalized Hölder spaces $H_0^\omega(\rho)$ via the indices m_ω and M_ω. *Fract. Calc. Appl. Anal.*, 7(4):437–458, 2004.
3. L. Maligranda. Indices and interpolation. *Dissertationes Math. (Rozprawy Mat.)*, 234:49, 1985.

The Approximation of Value Function for two Person Differential Games with Zero Sum

Ścibór Sobieski

Faculty of Mathematics and Computer Science, University of Lodz

INTRODUCTION

We consider the following optimal problem of Bolza,

$$\min_{u}\max_{z} J(x,u,z) = \int_{t_0}^{T} L(t,x(t),u(t),z(t))\,dt + l(x(T)). \tag{1}$$

over all absolutely continuous functions $x : [t_0, T] \to R^n$, $t_0 \geq 0$

$$\dot{x}(t) = f(t,x(t),u(t),z(t)), \quad \text{a.e. in } [t_0,T] \tag{2}$$

with the initial condition

$$x(t_0) = x_0, \quad x_0 \in R^n, \tag{3}$$

and

$$u(t) \in U, \quad t \in [t_0,T], \tag{4}$$

$$z(t) \in Z, \quad t \in [t_0,T], \tag{5}$$

where control functions $u(\cdot)$, $z(\cdot)$ are Lebesgue's measurable. $U \subset R^r$ and $Z \subset R^s$, $f : [0,T] \times R^n \times U \times Z \to R^n$, $L : [0,T] \times R^n \times U \times Z \to R$, $l : R^n \to R$.

We will assume

$$(H1) \begin{cases} a) L(t,\cdot,u,z),\, f(t,\cdot,u,z) \text{ satisfy a local Lipschitz condition and} \\ \quad L(\cdot,x,\cdot,\cdot),\, f(\cdot,x,\cdot,\cdot) \text{ are } L \times B-\text{measurable,} \\ b) l \text{ is lower semicontinuous in } \mathbb{R}^n. \end{cases}$$

THE VALUE FUNCTION AND ε-VALUES FUNCTIONS

Definition 1 *The functions $x(\cdot)$, $u(\cdot)$, $z(\cdot)$ are called admissible if they satisfy (2), (4), (5) and if $t \to L(t,x(t),u(t),z(t))$ is summable. The suitable function $t \to x(t)$ will be called admissible trajectory, and $t \to u(t)$, $t \to z(t)$ are then called admissible controls.*

Let us denote by $Y \subset [t_0,T] \times R^n$ the set covered by the graphs of all admissible trajectories, and assume that Y is a set with nonempty interior. So for any $(t_0,x_0) \in Y$ there exists an admissible function $x(t)$, $t \in [t_0,T]$, such that $x(t_0) = x_0$ and $(t,x(t)) \in Y$ for $t \in [t_0,T]$.

For our type of approximation we need stronger assumptions:

$$(H2) \begin{cases} a) L,\, f \text{ are Lipschitz functions and} \\ \quad Y \text{ is compact} \\ b) l \text{ is Lipschitz function.} \end{cases}$$

CP1046, *Selected Papers from ICNAAM 2007 and ICCMSE 2007*
edited by T. E. Simos, G. Maroulis, G. Psihoyios, and Ch. Tsitouras
© 2008 American Institute of Physics 978-0-7354-0574-5/08/$23.00

Definition 2 *Function* $(t,x) \to S(t,x)$, $(t,x) \in Y$, *is called a value function, if*

$$S(t,x) = \inf\sup\left\{\int_t^T L(s,x(s),u(s),z(s))\,ds + l(x(T))\right\}, \qquad (6)$$

where $\inf$ *is taken over all controls* $u(t) \in U$ *and* $\sup$ *over all controls* $z(t) \in Z$, $t \in [t_0,T]$, $x(t) = x$.

Definition 3 *The function* $(t,x) \to S_{\varepsilon l}(t,x)$, $(t,x) \in Y$ *is called lowewr ε-value function, and the function* $(t,x) \to S_{\varepsilon_h}(t,x)$, $(t,x) \in Y$ *is called upper ε-value function, if these functions satisfy the following inequalities*

$$S(t,x) - \varepsilon(T - t_0) \le S_{\varepsilon_l}(t,x) \le S_{\varepsilon_h}(t,x) \le S(t,x) + \varepsilon(T - t_0), \qquad (7)$$

where $(t,x) \in Y$, $(t,x) \to S(t,x)$ *is a value function, and* $\varepsilon > 0$ *is arbitrary fixed (see [2])*.

Of course there are infinitely many ε-values functions for a fixed $\varepsilon > 0$. And if $\varepsilon = 0$, the function $S_{\varepsilon_l}(\cdot,\cdot)$ and $S_{\varepsilon_h}(\cdot,\cdot)$ from Definition 2 define the value function.

Let us note that the functions $(t,x) \to S_{\varepsilon_l}(t,x)$ and $(t,x) \to S_{\varepsilon_h}(t,x)$ are finite.

APPROXIMATION

To construct an approximation of values functions we use the method of Hamilton-Jacobi inequalities (compare [1]). Let us take an arbitrary chosen C^1 functions $w_l(t,x)$ and $w_h(t,x)$, $w_l(t,x) \le w_h(t,x)$, $(t,x) \in Y$ with boundary conditions: $w_l(T,x) = l(x)$, $w_h(T,x) = l(x)$, $(T,x) \in Y$.

On the set Y we define functions $(t,x) \to F_l(t,x)$, $(t,x) \to F_h(t,x)$ corresponding to the right hand side of Hamilton-Jacobi equation:

$$F_l(t,x) := \frac{\partial}{\partial t}w_l(t,x) + \max_{z \in Z}\min_{u \in U}\left\{\frac{\partial}{\partial x}w_l(t,x)f(t,x,u,z(t)) + L(t,x,u,z)\right\}, \qquad (8)$$

$$F_h(t,x) := \frac{\partial}{\partial t}w_h(t,x) + \min_{u \in U}\max_{z \in Z}\left\{\frac{\partial}{\partial x}w_h(t,x)f(t,x,u,z) + L(t,x,u,z)\right\}, \qquad (9)$$

Functions $(t,x) \to F_l(t,x)$ and $(t,x) \to F_h(t,x)$ are continuous functions on Y, moreover they are Lipschitz functions on this set. Because Y is compact the functions $F_l(\cdot,\cdot)$, $F_h(\cdot,\cdot)$ attain their supremum and infimum so we have for some k_l, k^l, k_h, k^h:

$$k_l \le F_l(t,x) \le k^l, k_h \le F_h(t,x) \le k^h, \text{for all } (t,x) \in Y.$$

Let us choose the arbitrary $\eta > 0$. Next we divide uniformly $[k_l, k^l]$ with points $\{y_j^\eta\}_{j \in \mathbb{Z}}$, where $\mathbb{Z} = \{1,...,n\}$, in the following way:

$$y_1^\eta = k_l, y_n^\eta = k^l,$$
$$y_{j+1}^\eta - y_j^\eta = \eta, \ j \in \mathbb{Z}.$$

Now, let us divide Y set on the following subset P_j^{η,w_l}, $j \in \mathbb{Z}$:

$$P_j^{\eta,w_l} := \left\{(t,x) \in Y : y_j < F_l(t,x) \le y_{j+1}\right\}, \ j \in \mathbb{Z}.$$

Of course for all $i,j \in \mathbb{Z}, i \ne j, P_i^{\eta,w_l} \cap P_j^{\eta,w_l} = \varnothing$, and $\bigcup_{j \in \mathbb{Z}} P_j^{\eta,w_l} = Y$.

Let us define in Y the function $\ulcorner^{\eta,w_l}(\cdot,\cdot)$ depending on the above function $w_l(\cdot,\cdot)$ and number η as following:

$$\ulcorner^{\eta,w_l}(t,x) := -y_{j+1}^\eta \text{ for } (t,x) \in P_j^{\eta,w_l}, \ j \in \mathbb{Z}. \qquad (10)$$

So we have:

$$-\eta \le F_l(t,x) + \ulcorner^{\eta,w_l}(t,x) \le 0, \text{ for } (t,x) \in Y.$$

Analogously, on base (9), we define the function $\ulcorner^{\eta,w_h}(\cdot,\cdot)$.

Let

$$\tilde{U} = \{u(\cdot) : u(t) \in U, t \in [t_0,b], u(\cdot) \text{ - measurable}\},$$
$$\tilde{Z} = \{z(\cdot) : z(t) \in Z, t \in [t_0,b], z(\cdot) \text{ - measurable}\},$$
$$\mathscr{X} = \left\{\begin{array}{l} x_u^z(t), \ t \in [t_0,b] : x(t), \ t \in [t_0,b] \text{ admissible trajectory} \\ \text{generate by } u(\cdot) \in \tilde{U}, z(\cdot) \in \tilde{Z} \text{ and } x(t_0) = x_0. \end{array}\right\}.$$

THE MAIN RESULT

Theorem 1 *(i) The value:*

$$w_l(t_0, x_0) + \inf_{u(\cdot) \in \tilde{U}} \sup_{z \in \tilde{Z}} \left\{ - \int_{t_0}^{T} \Gamma^{\eta, w_l}(t, x_u^z(t)) dt \right\},$$

where function $\Gamma^{\eta, w_l}(\cdot, \cdot)$ is defined in (10), is a lower ε-value.

(ii) The value:

$$w_h(t_0, x_0) + \sup_{z \in \tilde{Z}} \inf_{u(\cdot) \in \tilde{U}} \left\{ - \int_{t_0}^{T} \Gamma^{\eta, w_h}(t, x_u^z(t)) dt \right\},$$

where the function $\Gamma^{\eta, w_l}(\cdot, \cdot)$ is defined analogously to $\Gamma^{\eta, w_l}(\cdot, \cdot)$, is a upper ε-value.

REFERENCES

1. J. Pustelnik, *A method for constructing the ε-value functions for the Bolza problem of optimal control*, Intl. J. of Applied Mathematics and Computer Science, to appear in 2005, Vol.15, No.2
2. Ś. Sobieski, *Properties of ε-value functions for a two person differential game*, Optimization. (submitted)

The Construction of Lyapunov Function for the Input Systems of ODE Type

Ścibór Sobieski and Paulina Puścian

Faculty of Mathematics and Computer Science, University of Lodz

Keywords: Lyapunov function, Lyapunov stability, numerical approximation

INTRODUCTION

We consider the mathematical models of input systems, described by continuous-time, finite dimensional ordinary differential equations

$$\dot{x} = f(t, x, u), \tag{1}$$

where $t \geq 0$, $x = (x_1, \ldots, x_n) \in \mathbb{R}^n$, represents the state variables, $u = (u_1, \ldots, u_n) \in \mathbb{R}^m$ represents the input variables and $f = (f_1, \ldots, f_n) : [0, +\infty) \times \mathbb{R}^n \times \mathbb{R}^m \to \mathbb{R}^n$. The control u belongs to some subset $U \subset \mathbb{R}^m$ which we assume to be a compact set. About f we assume:

A1 $f(\cdot, x, u)$ is measurable for each (x, u),

A2 for each $t > 0$, $f(t, \cdot, \cdot)$ is continuous,

A3 the function $f(\cdot, \cdot, \cdot)$ is locally essentially bounded on $[0, +\infty) \times B_r \times U$ (B_r is a ball in $\mathbb{R}^n$ with center zero and radius $r > 0$), $f(t, 0, u) = 0$, $t > 0$, $u \in U$.

These assumptions ensure that (1) has at least one solution (at least locally) for each measurable control $u(t) \in U$.

The problem we want to address in the paper is stability in control theory. There two problems of interest in nonlinear control theory, namely the existence of Lyapunov functions for Lyapunov stable or Lagrange stable systems and the existence of internal stabilizers, which in general do not have a solution in the class of continuous functions. We shall deal with the first problem. It is well known that semi-continuous Lyapunov functions are easier to construct, but then, it becomes more difficult to check that they are non-increasing along the trajectories. In such a case often the nonsmooth analysis methods are used.

The aim of the paper is a construction (of numerical type) a Lyapunov function for problem (1). To be sure that a construction is possible we assume that the problem has a Lyapunov function $\tilde{V}(t, x)$ being Lipschitz continuous in $[0, +\infty) \times B_r$ (on existence of such functions see [1], [3]). We look for a real function $V(t, x)$ defined in $[0, +\infty) \times B_r$ and fulfilling the following properties:

(i) there exist continuous functions $a(0) = 0, b(0) = 0$ and nondecreasing such that

$$a(\|x\|) \leq V(t, x) \leq b(\|x\|) \text{ for } t \in [0, +\infty), x \in B_r$$

(ii) for each Caratheodory solution $x(\cdot)$ of (1) and each interval $I \subseteq [0, +\infty)$ one has

$$t_1, t_2 \in I, t_1 < t_2 \implies V(t_1, x(t_2)) \geq V(t_2, x(t_2))$$

provided that $x(\cdot)$ is defined on I and $x(t) \in B_r$ for $t \in I$.

CONTRUCTION OF THE APPROXIMATION

To construct a Lyapunov function we use method of Hamilton-Jacobi inequalities (compare [2]). Let $Y = [0, +\infty) \times B_r$. Let us choose an arbitrary $C^1(Y)$ positive function $w(t, x)$ with bounded derivatives $\frac{\partial}{\partial t} w(t, x)$, $\frac{\partial}{\partial x} w(t, x)$ and boundary condition: $w(t, 0) = 0$, $t > 0$.

CP1046, *Selected Papers from ICNAAM 2007 and ICCMSE 2007*
edited by T. E. Simos, G. Maroulis, G. Psihoyios, and Ch. Tsitouras
© 2008 American Institute of Physics 978-0-7354-0574-5/08/$23.00

Next define function $Y \ni (t,x) \to F(t,x)$ corresponding to the right side of Hamilton-Jacobi equation:

$$F(t,x) := \frac{\partial}{\partial t}w(t,x) + \sup_{u \in U}\left\{\frac{\partial}{\partial x}w(t,x)f(t,x,u)\right\}. \tag{2}$$

Function $(t,x) \to F(t,x)$ is bounded on Y. So the infimum and supremum of $F(\cdot,\cdot)$ are finite and we have for same k_l, k_h that:

$$k_l \leq F(t,x) \leq k_h, \text{for all } (t,x) \in Y.$$

Let us choose an arbitrary $\eta > 0$. Next we divide uniformly the interval $[k_l,k_h]$ by points $k_l = y_1^\eta < y_2^\eta < ... < y_n^\eta = k_h$, such that:

$$y_{j+1}^\eta - y_j^\eta = \eta, \ j \in \{1,...,n\}.$$

Now, let us divide Y on the following subsets $P_j^{\eta,w}$, $j \in \{1,...,n\}$:

$$P_j^{\eta,w} := \left\{(t,x) \in Y : y_j < F(t,x) \leq y_{j+1}\right\}, \ j \in \{1,...,n\}.$$

Of course, for all $i,j \in \{1,...,n\}, i \neq j, P_i^{\eta,w} \cap P_j^{\eta,w} = \varnothing$, and $\bigcup_{j\in\mathbb{Z}}P_j^{\eta,w} = Y$.

Let us define in Y a function $\mathfrak{l}^{\eta,w}(\cdot,\cdot)$ depending on function $w(\cdot,\cdot)$ and number η as following:

$$\mathfrak{l}^{\eta,w}(t,x) \quad : \quad = -y_{j+1} \text{ for } (t,x) \in P_j^{\eta,w}, \ j \in \{1,...,n\} \tag{3}$$

$$\mathfrak{l}^{\eta,w}(t,0) \quad = \quad 0, t > 0. \tag{4}$$

So we have:

$$\forall_{(t,x)\in Y} \ -\eta \leq F(t,x) + \mathfrak{l}^{\eta,w}(t,x) \leq 0$$

and thus, for each $(t,x) \in Y$ and each $u \in U$

$$\frac{\partial}{\partial t}w(t,x) + \frac{\partial}{\partial x}w(t,x)f(t,x,u) + \mathfrak{l}^{\eta,w}(t,x) \leq 0.$$

THE MAIN RESULT

Theorem 1 *The function*

$$w(t,x) + t\mathfrak{l}^{\eta,w}(t,x)$$

satisfies (i) and (ii), where the function $\mathfrak{l}^{\eta,w}(\cdot,\cdot)$ is defined in (3).

REFERENCES

1. A. Bacciotti and L. Rosier, *Liapunov Function and Stability in Control Theory*, Springer-Verlag, Berlin, Heidelberg, 2005, 2 edn., ISBN 3-540-21332-5.
2. J. Pustelnik, *A method for constructing the ε-value functions for the Bolza problem of optimal control*, Intl. J. of Applied Mathematics and Computer Science, to appear in 2005, Vol.15, No.2
3. M. Grzanek, A. Michalak, A. Rogowski, *A nonsmooth Lyapunov function and stability for ODE's of Caratheodory type*, Nonlinear Analysis, 2007, http://dx.doi.org/10.1016/j.na.2007.05.022.

A Different Type of Hardy-Hilbert's Integral Inequality

W. T. Sulaiman

Abstract. We give new kind of Hardy-Hilbert's integral inequality via homogeneous functions as well as some other generalization. Special cases are also obtained.

Keywords: Hardy-Hilbert's inequality, Holder's inequality, Homogeneous function, Beta function.

MSC: 26D15

INTRODUCTION

Let $f, g \geq 0$ satisfy

$$0 < \int_0^\infty f^2(t)\,dt < \infty \quad and \quad 0 < \int_0^\infty g^2(t)\,dt < \infty,$$

then

$$\int_0^\infty \int_0^\infty \frac{f(x)g(y)}{x+y}\,dx\,dy < \pi\left(\int_0^\infty f^2(t)\,dt \int_0^\infty g^2(t)\,dt\right)^{1/2}, \tag{1}$$

where the constant factor π is the best possible (cf. Hardy et al. [2]).Inequality (1) is well known as Hilbert's integral inequality. This inequality had been extended by
Hardy [1] as follows

If p>1, $\frac{1}{p}+\frac{1}{q}=1$, $f, g \geq 0$ satisfy

$$0 < \int_0^\infty f^p(t)\,dt < \infty \quad and \quad \int_0^\infty g^q(t)\,dt < \infty,$$

then

$$\int_0^\infty \int_0^\infty \frac{f(x)g(y)}{x+y}\,dx\,dy < \frac{\pi}{\sin(\pi/p)}\left(\int_0^\infty f^p(t)\,dt\right)^{1/p}\left(\int_0^\infty g^q(t)\,dt\right)^{1/q}, \tag{2}$$

where the constant factor $\dfrac{\pi}{\sin(\pi/p)}$ is the best possible. Inequality (2) is called

Hardy-Hilbert's integral inequality and is important in analysis and application (cf. Mitrinovic et al. [3]).

B. Yang gave the following extension of (2) as follows:

Theorem [4]. *If* $\lambda > 2 - \min\{p, q\}$, $f, g \geq 0$, satisfy

$$0 < \int_0^\infty t^{1-\lambda} f^p(t)\,dt < \infty \quad and \quad \int_0^\infty t^{1-\lambda} g^q(t)\,dt < \infty,$$

then

CP1046, *Selected Papers from ICNAAM 2007 and ICCMSE 2007*
edited by T. E. Simos, G. Maroulis, G. Psihoyios, and Ch. Tsitouras
© 2008 American Institute of Physics 978-0-7354-0574-5/08/$23.00

$$\int_0^\infty \int_0^\infty \frac{f(x)g(y)}{(x+y)^\lambda} dx\, dy < k_\lambda(p) \left(\int_0^\infty t^{1-\lambda} f^p(t)\, dt \right)^{1/p} \left(\int_0^\infty t^{1-\lambda} g^q(t)\, dt \right)^{1/q}, \tag{3}$$

where the constant factor $k_\lambda(p) = B\left(\frac{p+\lambda-2}{p}, \frac{q+\lambda-2}{q}\right)$ is the best possible, B is the beta function. The function $f(x,y)$ is said to be homogeneous of degree λ, if

$$f(tx, ty) = t^\lambda f(x, y), \ t > 0. \tag{4}$$

The object of this paper is that to give some new inequalities similar to that of Hardy-Hilbert's inequality.

NEW RESULTS

We state and prove the following:

Theorem 1. Let $f, g \ge 0$, h is positive function of two variables, and homogeneous of degree $\lambda > 0$, $p > 1$, $\frac{1}{p} + \frac{1}{q} = 1$. T>0. Then

$$\int_0^T \int_0^T \frac{f(u)g(v)}{h(u,v)} du\, dv \le \frac{1}{p} \int_0^T u^{\alpha(2-p)+1-\lambda} f^p(u) K_1(u)\, du \tag{5}$$

$$+ \frac{1}{q} \int_0^T v^{2(2-q)+1-\lambda} g^q(v) K_2(v)\, dv,$$

where

$$K_1(u) = \int_0^{T/u} \frac{y^\alpha}{h(1,y)} dy, \quad K_2(v) = \int_0^{T/v} \frac{x^\alpha}{h(x,1)} dx,$$

and

$$\int_0^T v^{\frac{\alpha(2-q)+1-\lambda}{1-q}} K_2^{\frac{1}{1-q}}(v) \left(\int_0^T \frac{f(u)}{h(u,v)} du \right)^p dv \le \int_0^T u^{\alpha(2-p)+1-\lambda} f^p(u) K_1(u)\, du. \tag{6}$$

Inequalities (5) and (6) are equivalent.
In particular, for $T = \infty$, $\alpha = \lambda/2 - 1$, we have

$$\int_0^\infty \int_0^\infty \frac{f(u)g(v)}{h(u,v)} du\, dv \le K \left(\frac{1}{p} \int_0^\infty u^{\left(1-\frac{\lambda}{2}\right)p-1} f^p(u)\, du + \frac{1}{q} \int_0^\infty v^{\left(1-\frac{\lambda}{2}\right)q-1} g^q v)\, dv \right) \tag{7}$$

where

$$K = \int_0^T \frac{z^{\frac{\lambda}{2}-1}}{h(1,z)} dz < q,$$

and

$$\int_0^\infty v^{\frac{\left(1-\frac{\lambda}{2}\right)q-1}{1-q}} \left(\int_0^\infty \frac{f(u)}{h(u,v)} du \right)^p dv \le \frac{K/p}{1-K/q} \int_0^\infty u^{\left(1-\frac{\lambda}{2}\right)p-1} f^p(u)\, du. \tag{8}$$

The inequalities (7) and (8) are equivalent.
Proof. Making use of the inequality

$$ab \le \frac{a^p}{p} + \frac{b^q}{q}, \quad a,b \ge 0, \quad p > 1, \quad \tfrac{1}{p} + \tfrac{1}{q} = 1,$$

$$\int_0^T\int_0^T \frac{f(u)\,g(v)}{h(u,v)}\,du\,dv = \int_0^T\int_0^T \frac{u^{-\frac{\alpha}{q}} v^{\frac{\alpha}{p}} f^p(u)}{h^{\frac{1}{p}}(u,v)} \times \frac{u^{\frac{\alpha}{q}} v^{-\frac{\alpha}{p}} g^q(v)}{h^{\frac{1}{q}}(u,v)}\,du\,dv$$

$$\le \frac{1}{p}\int_0^T\int_0^T \frac{u^{(1-p)\alpha} v^{\alpha} f^p(u)}{h(u,v)}\,du\,dv \;+\; \frac{1}{q}\int_0^T\int_0^T \frac{v^{(1-q)\alpha} u^{\alpha} g^q(v)}{h(u,v)}\,du\,dv$$

$$= \frac{M}{p} + \frac{N}{q}.$$

$$M = \int_0^T u^{(1-p)\alpha} f^p(u)\,du \int_0^T \frac{v^{\alpha}}{h(u,v)}\,dv.$$

Observe that on putting $v = uy$, $dv = u\,dy$, $0 \le y \le T/u$, we have

$$\int_0^T \frac{v^{\alpha}}{h(u,v)}\,dv = \int_0^{T/u} \frac{(uy)^{\alpha} u}{h(u,uy)}\,dy = u^{1+\alpha-\lambda}\int_0^{T/u} \frac{y^{\alpha}}{h(1,y)}\,dy,$$

which implies

$$M = \int_0^T u^{(2-p)\alpha+1-\lambda} f^p(u)\,K_1(u)\,du.$$

$$N = \int_0^T v^{(1-q)\alpha} g^q(v)\,dv \int_0^T \frac{u^{\alpha}}{h(u,v)}\,du.$$

Now, on putting $u = vx$, $du = v\,dx$, $0 \le x \le T/v$, we have

$$N = \int_0^T v^{(2-q)+1-\lambda\alpha} g^q(v)\,dv \int_0^{T/v} \frac{x^{\alpha}}{h(x,1)}\,dx$$

$$= \int_0^T v^{(2-q)\alpha+1-\lambda} g^q(v)\,K_2(v)\,dv.$$

Therefore (5) is satisfied. To prove the equivalence of (5) and (6), let (6) be satisfied. Then, we have

$$\int_0^T v^{\frac{(2-q)\alpha+1-\lambda}{1-q}} K_2^{\frac{1}{1-q}}(v)\left(\int_0^T \frac{f(u)}{h(u,v)}\,du\right)^p dv$$

$$= \int_0^T\int_0^T \frac{f(u)\, v^{\frac{(2-q)\alpha+1-\lambda}{1-q}} K_2^{\frac{1}{1-q}}(v)\left(\int_0^T \frac{f(u)}{h(u,v)}\,du\right)^{p-1}}{h(u,v)}\,du\,dv$$

$$\le \frac{1}{p}\int_0^T u^{(2-p)\alpha+1-\lambda} f^p(u)\,K_1(u)\,du$$

$$+ \frac{1}{q}\int_0^T v^{(2-q)\alpha+1-\lambda}\, v^{\frac{q}{1-q}((2-q)\alpha+1-\lambda)}\, K_2(v)\,K_2^{\frac{q}{1-q}}(v)\left(\int_0^T \frac{f(u)}{h(u,v)}\,du\right)^{q(p-1)} dv$$

$$= \frac{1}{p} \int_0^T u^{(2-p)\alpha+1-\lambda} f^p(u) K_1(u)\, du$$

$$+ \frac{1}{q} \int_0^T v^{\frac{(2-q)\alpha+1-\lambda}{1-q}} K_2^{\frac{1}{1-q}}(v) \left(\int_0^T \frac{f(u)}{h(u,v)}\, du \right)^p dv.$$

This implies

$$\int_0^T v^{\frac{(2-q)\alpha+1-\lambda}{1-q}} K_2^{\frac{1}{1-q}}(v) \left(\int_0^T \frac{f(u)}{h(u,v)}\, du \right)^p dv \le \int_0^T u^{(2-p)\alpha+1-\lambda} f^p(u) K_1(u)\, du.$$

Now, Let (5) be satisfied, we have

$$\int_0^T \int_0^T \frac{f(u)\, g(v)}{h(u,v)}\, du\, dv$$

$$= \int_0^T v^{-\frac{(2-q)\alpha+1-\lambda}{(1-q)p}} g(v) K_2^{\frac{1}{q}}(v)\; v^{\frac{(2-q)\alpha+1-\lambda}{(1-q)p}} K_2^{-\frac{1}{q}}(v) \left(\int_0^T \frac{f(u)}{h(u,v)}\, du \right) dv$$

$$\le \frac{1}{q} \int_0^T v^{(2-q)\alpha+1-\lambda} g^q(v) K_2(v)\, dv$$

$$+ \frac{1}{p} \int_0^T v^{\frac{(2-q)\alpha+1-\lambda}{1-q}} K_2^{\frac{1}{1-q}}(v) \left(\int_0^T \frac{f(u)}{h(u,v)}\, du \right)^p dv$$

$$\le \frac{1}{q} \int_0^T v^{(2-q)\alpha+1-\lambda} g^q(v) K_2(v)\, dv + \frac{1}{p} \int_0^T u^{(2-p)\alpha+1-\lambda} f^p(u) K_1(u)\, du.$$

In the particular case, (7) and (8) follows from (5) and (6) respectively, noticing that for $\alpha = \lambda/2 - 1$, $T = \infty$,

$$K_1 = K_2 = K = \int_0^\infty \frac{z^{\frac{\lambda}{2}-1}}{h(1,z)}\, dz,$$

follows as

$$\int_0^\infty \frac{x^{\frac{\lambda}{2}-1}}{h(x,1)}\, dx = \int_0^\infty \frac{x^{\frac{\lambda}{2}-1}}{h(x,xx^{-1})}\, dx = \int_0^\infty \frac{x^{-\frac{\lambda}{2}-1}}{h(1,x^{-1})}\, dx = \int_0^\infty \frac{z^{\frac{\lambda}{2}-1}}{h(1,z)}\, dz.$$

This completes the proof of the theorem.

The following lemma is needed for the coming result

Lemma 2. Let $a_i \ge 0$, $p_i > 1$, $i = 1,\ldots,n$, $\sum_{i=1}^n \frac{1}{p_i} = 1$. Then

$$\prod_{i=1}^n a_i \le \sum_{i=1}^n \frac{a_i^{p_i}}{p_i}.$$

Proof. We shall prove the lemma using induction. Obviously it is true for n=1,2. Suppose it is true for n-1. Since

$$\frac{1}{1/\sum_{i=1}^{n-1}(1/p_i)} + \frac{1}{p_n} = 1, \quad \sum_{i=1}^{n-1}\left(\frac{1}{p_i\left(\sum_{i=1}^{n-1} 1/p_i\right)} \right) = 1,$$

then

$$\prod_{i=1}^{n} a_i = \left(\prod_{i=1}^{n-1} a_i\right) a_n \le \left(\sum_{i=1}^{n-1} 1/p_i\right)\left(\prod_{i=1}^{n-1} a_i\right)^{1/\left(\sum_{i=1}^{n-1} 1/p_i\right)} + \frac{1}{p_n} a_n^{p_n}$$

$$= \left(\sum_{i=1}^{n-1} 1/p_i\right)\left(\prod_{i=1}^{n-1} a_i^{1/\left(\sum_{i=1}^{n-1} 1/p_i\right)}\right) + \frac{1}{p_n} a_n^{p_n}$$

$$\le \left(\sum_{i=1}^{n-1} 1/p_i\right) \sum_{i=1}^{n-1} \frac{1}{\left(\sum_{i=1}^{n-1} 1/p_i\right) p_i} \left(a_i^{1/\left(\sum_{i=1}^{n-1} 1/p_i\right)}\right)^{\left(\sum_{i=1}^{n-1} 1/p_i\right) p_i}$$

$$= \sum_{i=1}^{n-1} \frac{1}{p_i} a_i^{p_i} + \frac{1}{P_n} a_n^{p_n} = \sum_{i=1}^{n} \frac{a_i^{p_i}}{p_i}.$$

Theorem 3. Let $f_i \ge 0$, $p_i > 1$, $\lambda > p_i - 1$, $i = 1,...,n$, $\sum_{i=1}^{n} \frac{1}{p_i} = 1$. Then

$$\int_0^\infty ... \int_0^\infty \frac{f_1(t_1)...f_n(t_n)}{(t_1 + ... + t_n)^\lambda} dt_1...dt_n \le \frac{1}{\Gamma(\lambda)} \sum_{i=1}^{n} \frac{1}{p_i} \Gamma^{p_i}\left(\frac{\lambda}{p_i}\right) \int_0^\infty t^{p_i - \lambda - 2} f_i^{p_i}(t) dt. \qquad (9)$$

Proof. Define, for i=1,...,n

$$F_i(x) = \int_0^\infty e^{-tx} f_i(t) dt.$$

Observe that, via lemma 2,

$$\int_0^\infty s^{\lambda-1} F_1(s)...F_n(s) ds$$

$$= \int_0^\infty s^{\frac{\lambda-1}{p_1}} F_1(s)...s^{\frac{\lambda-1}{p_n}} F_n(s) ds$$

$$\le \int_0^\infty \sum_{i=1}^{n} \frac{s^{\lambda-1} F_i^{p_i}(s)}{p_i} ds$$

$$= \sum_{i=1}^{n} \frac{1}{p_i} \int_0^\infty s^{\lambda-1} F_i^{p_i}(s) ds.$$

Since

$$F_i^{p_i}(s) = \left(\int_0^\infty e^{-ts} f_i(t) dt\right)^{p_i}$$

$$\le \int_0^\infty e^{-ts} f_i^{p_i}(t) dt \left(\int_0^\infty e^{-ts} dt\right)^{p_i-1}$$

$$= s^{1-p_i} \int_0^\infty e^{-ts} f_i^{p_i}(t) dt,$$

we obtain

$$\int_0^\infty s^{\lambda-1} F_1(s)...F_n(s) ds \le \sum_{i=1}^{n} \frac{1}{p_i} \int_0^\infty s^{\lambda-p_i} \int_0^\infty e^{-ts} f_i^{p_i}(t) dt\, ds$$

$$= \sum_{i=1}^{n} \frac{1}{p_i} \int_0^\infty f_i^{p_i}(t)\, dt \int_0^\infty s^{\lambda - p_i}\, e^{-ts}\, ds$$

$$= \sum_{i=1}^{n} \frac{1}{p_i} \int_0^\infty t^{p_i - \lambda - 1} f_i^{p_i}(t)\, dt \int_0^\infty u^{\lambda - p_i} e^{-u}\, du$$

$$= \sum_{i=1}^{n} \frac{\Gamma(1 + \lambda - p_i)}{p_i} \int_0^\infty t^{p_i - \lambda - 1} f_i^{p_i}(t)\, dt.$$

On the other hand,

$$\int_0^\infty s^{\lambda - 1} F_1(s)...F_n(s)\, ds$$

$$= \int_0^\infty s^{\lambda - 1} \int_0^\infty e^{-s t_1} f_1(t_1)\, dt_1 ... \int_0^\infty e^{-s t_n} f_n(t_n)\, dt_n$$

$$= \int_0^\infty ... \int_0^\infty f_1(t_1)...f_n(t_n)\, dt_1...dt_n \int_0^\infty s^{\lambda - 1} e^{-s(t_1 + ... + t_n)}\, ds$$

$$= \int_0^\infty ... \int_0^\infty \frac{f_1(t_1)...f_n(t_n)}{(t_1 + ... + t_n)^\lambda}\, dt_1...dt_n \int_0^\infty z^{\lambda - 1} e^{-z}\, dz$$

$$= \Gamma(\lambda) \int_0^\infty ... \int_0^\infty \frac{f_1(t_1)...f_n(t_n)}{(t_1 + ... + t_n)^\lambda}\, dt_1...dt_n.$$

Summarizing, we have

$$\int_0^\infty ... \int_0^\infty \frac{f_1(t_1)...f_n(t_n)}{(t_1 + ... + t_n)^\lambda}\, dt_1...dt_n \le \frac{1}{\Gamma(\lambda)} \sum_{i=1}^{n} \frac{\Gamma(1 + \lambda - p_i)}{p_i} \int_0^\infty t^{p_i - \lambda - 1} f_i^{p_i}(t)\, dt.$$

APPLICATIONS

Corollary 4. On putting $h(x, y) = (x + y)^\lambda$, which is homogeneous of degree λ, in theorem 1, inequality (7), we obtain

$$\int_0^\infty \int_0^\infty \frac{f(x)\, g(y)}{(x + y)^\lambda}\, dx\, dy \le B\left(\tfrac{\lambda}{2}, \tfrac{\lambda}{2}\right) \left(\frac{1}{p} \int_0^\infty x^{\left(1 - \frac{\lambda}{2}\right) p - 1} f^p(x)\, dx + \int_0^\infty y^{\left(1 - \frac{\lambda}{2}\right) q - 1} g^q(y)\, dy \right).$$

Corollary 5. On putting $f(x, y) = x^\lambda + y^\lambda$, which is homogeneous of degree λ, in theorem 1, inequality (7), we obtain

$$\int_0^\infty \int_0^\infty \frac{f(x)\, g(y)}{x^\lambda + y^\lambda}\, dx\, dy \le \frac{\pi}{\lambda} \left(\frac{1}{p} \int_0^\infty x^{\left(1 - \frac{\lambda}{2}\right) p - 1} f^p(x)\, dx + \frac{1}{q} \int_0^\infty y^{\left(1 - \frac{\lambda}{2}\right) q - 1} g^q(y)\, dy \right).$$

Corollary 6. On putting $f(x, y) = \left(x^{\sqrt{\lambda}} + y^{\sqrt{\lambda}}\right)^{\sqrt{\lambda}}$, which is homogeneous of degree λ, in theorem 1, inequality (7), we get

$$\int_0^\infty \int_0^\infty \frac{f(x)\,g(y)}{\left(x^{\sqrt{\lambda}} + y^{\sqrt{\lambda}}\right)^{\sqrt{\lambda}}}\,dx\,dy \leq \frac{1}{\sqrt{\lambda}}\,B\left(\frac{\sqrt{\lambda}}{2}, \frac{\sqrt{\lambda}}{2}\right)\left(\frac{1}{p}\int_0^\infty x^{\left(1-\frac{\lambda}{2}\right)p-1} f^p(x)\,dx + \frac{1}{q}\int_0^\infty y^{\left(1-\frac{\lambda}{2}\right)q-1} g^q(y)\,dy\right).$$

CONCLUSION

In this paper, we have changed the normal way to presenting inequalities similar to Hardy-Hilbert's integral inequality, by using the inequality $ab \leq \dfrac{a^p}{p} + \dfrac{b^q}{q}$, $a, b > 0$, $p > 1$, $\frac{1}{p} + \frac{1}{q} = 1$.

This leads us to have the R.H.S. of these inequalities of the Hardy-Hilbert's type containing sums instead of products. As well as we have the weight functions represented as homogeneous functions in order to generalize this kind of functions (see the corollaries). Finally a short method (in Theorem 3) gives a generalization to n variables for such inequalities.

REFERENCES

[1] G. H. Hardy, Note on a theorem of Hilbert concerning series of positive terms, Proc. Math. Soc. 23(2) (1925), Records of Proc. XLV-XLVI.
[2] G. H. Hardy, J. E. Littlewood and G. Polya, Inequalities, Cambridge University Press, Cambridge, 1952.
[3] D. S. Mitrinovic, J. E. Pecaric and A. M. Fink, Inequalities involving functions and their integrals and derivatives, Kluwer Academic Publishers, Bosten, 1991.
[4] B. Yang, On Hardy-Hilbert's integral inequality, J. Math. Anal. Appl. 261 (2001) 295-306.

Fast and Efficient Solution of Scattering Integral Equations

Miroslav Šulc, Přemysl Kolorenč, Michal Tarana, and Jiří Horáček

Faculty of Mathematics and Physics, Charles University in Prague, V Holešovičkách 2, 180 00 Praha 8, Czech Republic

Abstract. In this work we propose a method for numerical treatment of integral equations describing scattering of low-energy electrons with atoms and molecules. The method is based on a combination of R-matrix approach with the Schwinger-Lanczos method proposed by the authors. It is shown on the example of scattering of electrons by hydrogen atoms in the static exchange approximation that the method is very fast, economic and very accurate. By using only 64 meshpoints the accuracy of 9 significant figures can be easily obtained.

Keywords: Numerical solution of integral equations, electron atom collisions
PACS: 02.60Nm, 02.70.-c, 34.80.Bm, 34.80.-i

INTRODUCTION

The scattering of electrons with atoms and molecules is in principle described by appropriate Schrödinger equation. This equation however contains "nonlocal", this means integral, terms. As a consequence the Schrödinger equation is no longer a differential equation but an integrodifferential equation. Because of the special boundary conditions describing scattering processes it is much more convenient to transform this integrodifferential equation into an Fredholm integral equation of the second type the so called Lipmann-Schwinger equation [1]. In the Dirac notation used in physics we can formulate it in the following way

$$|\phi\rangle = |u\rangle + G_0(E)(V+W)|\phi\rangle, \tag{1}$$

where $|u\rangle$ denotes the incident plane wave, $G_0(E)$ is the Green's function of the free particle's Hamiltonian defined as

$$G_0(E) = (E - H_0)^{-1},$$

V resp. W stands for the local resp. nonlocal part of the interaction under consideration and finally $|\phi\rangle$ represents the sought solution. Simple algebraic manipulations furnish another insight onto this equation. Explicitly written

$$|\phi\rangle = (1 - G_0(E)V)^{-1}|u\rangle + (1 - G_0(E)V)^{-1}G_0(E)W|\phi\rangle \equiv |\overline{u}\rangle + G(E)W|\phi\rangle, \tag{2}$$

where we have introduced the symbol $|\overline{u}\rangle$ for the "distorted" wave obeying an equation of the same structure as Eq. (1). Namely

$$|\overline{u}\rangle = |u\rangle + G_0(E)V|\overline{u}\rangle. \tag{3}$$

After solving Eq. (2), the quantity of principal interest is the matrix element $\langle u|V+W|\phi\rangle$, which is directly proportional to the tangent of the phase shift (proper normalisation condition on $|u\rangle$ is to be imposed). Simple algebra reveals, that this matrix element can be equivalently computed as $\langle u|V|\overline{u}\rangle + \langle \overline{u}|W|\phi\rangle$. The latter approach is more useful than the former because the second summand is naturally rendered as a byproduct of the Schwinger-Lanczos iterative algorithm described below.

RLS ALGORITHM

The proposed algorithm for solving Eq.(1) is based on a combination of two very distinct ideas: 1. R-matrix approach [3] used to construct the Green's function corresponding to the local part of the interaction, 2. the Schwinger-Lanczos approach proposed by the authors [4, 5] used for the treatment of nonlocal operators. Hereafter we will refer to this algorithm as the RLS algorithm.

CP1046, *Selected Papers from ICNAAM 2007 and ICCMSE 2007*
edited by T. E. Simos, G. Maroulis, G. Psihoyios, and Ch. Tsitouras
© 2008 American Institute of Physics 978-0-7354-0574-5/08/$23.00

The Green's function for the local part of the interaction V is taken in the form [3]

$$G(r,r',E) = \sum_n \frac{\Psi_n(r)\Psi_n(r')}{(E_n - E)} + \frac{A}{2(1-A\mathscr{R})} \sum_{nm} \frac{\Psi_n(r_f)\Psi_m(r_f)\Psi_n(r)\Psi_m(r')}{(E_n - E)(E_m - E)}, \tag{4}$$

where

$$\mathscr{R} = \frac{1}{2} \sum_n \frac{\Psi_n(r_f)\Psi_n(r_f)}{(E_n - E)}, \tag{5}$$

$\Psi_n(r)$ are solutions of the R-matrix eigen-value problem

$$\tilde{K}\Psi_n(r) + \hat{V}\Psi_n(r) = E_n\Psi_n(r) \tag{6}$$

and A denotes the logarithmic derivative of the final solution at r_f. Here $\tilde{K}$ is the symmetrized kinetic energy operator

$$\tilde{K}_{ij} = \frac{1}{2} \int_0^{r_f} \frac{d\Psi_i}{dx} \frac{d\Psi j}{dx}, \tag{7}$$

where r_f (being presumably asymptotically large) represents the integration range.

The Schwinger-Lanczos method is based on the construction of Krylov basis set of vectors $|g_i\rangle$

$$\langle g_k | W | g_l \rangle = \delta_{kl} \tag{8}$$

chosen such that matrix WG_0W is tridiagonal

$$\langle g_{k-1} | WG_0W | g_k \rangle = \langle g_k | WG_0W | g_{k-1} \rangle = \beta_{k-1}, \tag{9}$$
$$\langle g_k | WG_0W | g_k \rangle = \alpha_k, \tag{10}$$
$$\langle g_k | WG_0W | g_l \rangle = 0 \quad \text{for} \quad |k-l| \geq 2 \tag{11}$$

where W is the nonlocal part of the interaction [4]. This approach is very general and poses no restriction on the form of the nonlocality. This algorithm yields directly the nonlocal contribution to the final T-matrix in the form of a continued fraction [5]

$$T^N = \langle \overline{u} | W | g_1 \rangle (M^{-1})_{11} \langle g_1 | W | \overline{u} \rangle = \cfrac{\langle \overline{u} | W | \overline{u} \rangle}{1 - \alpha_1 - \cfrac{\beta_1^2}{1 - \alpha_2 - \cfrac{\beta_2^2}{1 - \alpha_3 - \cdots - \cfrac{\beta_{N-1}^2}{1 - \alpha_N}}}}. \tag{12}$$

REFERENCE METHODS

In order to compare the numerical performance of the RLS approach with other methods we carried out calculations for four methods described in the literature:

- **S–IEM** is a spectral method for solving integral equation developed recently by G. H. Rawitscher et. al. Practical details can be found in [6, 7, 8].

- **M–IEM** is an iterative method originally introduced by B. T. Kim and T. Udagawa in [9]. Roughly speaking, the core of this method relies on utilising the Lanczos iterations for the complete LS equation.

- **N–IEM** stands for an older non iterative integral equation method based on the work by W. N. Sams and D. J. Kouri [10]. In principle, the final solution of the original one channel LS equation is expressed formally as a linear combination of two terms. The coefficients in this linear combination are profitably chosen (a linear set of two equations is to be solved to ensure this property) so that each of the two terms satisfies a Volterra integral equation of the second kind, which is then handled by standard means using e.g. trapezoidal quadrature rule.

- **MCFV** is an iterative method based on series of papers [11, 12, 13] by J. Horáček and T. Sasakawa. In this approach the local and nonlocal parts of the complete potential are also handled separately as in the RLS method. The Green's function is constructed in this case directly from the two independent solutions of the free Schrödinger equation. The implementation details can be found in [14].

NUMERICAL TESTS AND RESULTS

For the numerical test of the proposed algorithm we calculated the s-wave phase shift for scattering of electrons by ground-state hydrogen atoms in the static exchange approximation, both for the singlet $\delta^{(+)}$ and triplet $\delta^{(-)}$ state. We have chosen this model because similar calculations using different approaches (reference methods described above) can be found in the literature [7] offering the possibility to.compare the results.

For the s–wave, the scattering problem reduces to solving an integral equation for the radial part of the corresponding partial wave. Namely

$$\left(\frac{d^2}{dr^2}+k^2\right)R_0(r) = V(r)R_0(r) \pm \int_0^\infty K(r,r')R_0(r')\,dr' \tag{13}$$

where

$$V(r) = -2re^{-2r}\left(1+\frac{1}{r}\right) \tag{14}$$

$$K(r,r') = 2v(r)u(r') + \gamma u(r)u(r') \text{ for } r' < r \tag{15}$$

$$K(r,r') = 2u(r)v(r') + \gamma u(r)u(r') \text{ for } r' > r. \tag{16}$$

Here

$$u(r) = 2re^{-r}, \tag{17}$$

$$v(r) = \frac{u(r)}{r} = 2e^{-r} \tag{18}$$

and

$$\gamma = -k^2 - 1. \tag{19}$$

As we have already mentioned, our main goal is to compute the phase shift for the s–wave of the singlet as well as the triplet state. Since a relatively detailed study of the N–IEM method is found in [10], we tested the method presented in this paper for the same set of parameters. Results of the other mentioned methods for these parameters can be found in [7] and [14]. The case chosen is the singlet phase shift with exchange $\delta^{(+)}$. The wave number of the incident particle is $k = 0.2/a_0$, where a_0 is the Bohr radius ($a_0 = 1$ in atomic units), and the maximum radial distance (r_f, range of integration) is set to $20\,a_0$. As concerns the integration method used in the current implementation, we used the so called Clenshaw–Curtis quadrature, the details of which can be found e.g. in [2].

The results are summarised in Tab. 1 (typical number of sectors to obtain this accuracy was 4-5, while cca. 20 R-matrix basis functions per sector and 5-6 Lanczos iterations were needed), where the number of significant figures in the second column is determined from the stability of the phase shift value after rounding, when compared to the result corresponding to higher number of mesh points. It is seen that as few as 64 meshpoints yield the accuracy of at least 9 significant digits whereas the most recent approach of Rawitscher et al. [7] requires at least 80 digits to reach similar accuracy. Other methods require significantly larger number of meshpoints. Moreover, the RLS method is very advantageous in case when the calculation for more than one value of energy is required. The core of the calculation, the construction of the R-matrix basis, has to be done only once and then used for an arbitrary energy.

A more transparent overview of the convergence properties of these methods is depicted in Fig. (1), in which we are dealing with the "inverse" task, i.e. with determining the number of mesh points necessary to obtain a prescribed number of significant digits.

As can be seen from this picture, the proposed RLS method is very accurate even for very low number of mesh points. We believe, that it is primarily due to the accurate construction of the Green's function via the R–matrix approach. The problem of electron scattering with hydrogen atom in the static exchange approximation discussed here is in fact a very simple problem. The integral kernel is a real smooth non-oscillatory function with only one cusp. The problem of scattering of electrons with molecules is much more difficult. This will be discussed in a forthcoming paper.

ACKNOWLEDGMENTS

This work was supported by the grant GAUK No 257718 of the Charles University in Prague, by the Center of Theoretical Astrophysics LC06014 of the Ministry of Education, Youth and Sports of the Czech republic and by the Grant Agency of the Academy of Sciences of the Czech Republic grant no. IA100400501.

TABLE 1. Calculated phase shift $\delta^{(+)}$ for several reference methods. In the last column the number of mesh points used in the calculation is shown. Details are described below.

method	$\delta^{(+)}$	number of m.p.
RLS	1.87015788	64
S–IEM	1.8701579	80
MCFV	1.8701579	128
M–IEM	1.870156	4000
N–IEM	1.87015	4000

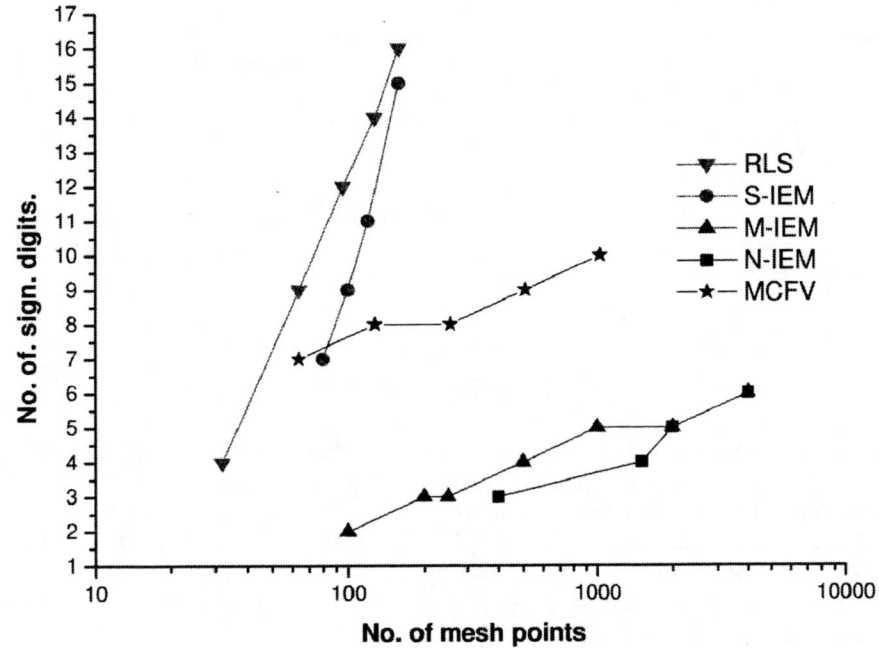

FIGURE 1. This figure shows the number of significant digits in the calculation of the triplet $\delta^{(+)}(k)$ phase shift as obtained by different methods.

REFERENCES

1. B. A. Lippmann, and J. Schwinger, *Phys. Rev.* **79**, 469–480 (1950).
2. S.-Y. Kang and I. Koltracht and G. Rawitscher, *Math. Comp.* **72**, 729–756 (2002)
3. G. V. Milnikov, J. Horáček, and H. Nakamura, *Comp. Phys. Comm.* **135**, 278–292 (2001).
4. H. D. Meyer, J. Horáček, and L. S. Cederbaum, *Phys. Rev.* **A43**, 3587–3596 (1991).
5. M. Čížek, J. Horáček, and H. D. Meyer, *Comp. Phys. Comm.* **131**, 41–51 (2000).
6. R. A. Gonzales, J. Eisert, I. Kolrath, and G. H. Rawitscher, *J. Comp. Phys.* **134**, 134–149 (1997).
7. G. H. Rawitscher, S.-Y. Kang, and I. Koltrath, *J. Chem. Phys.* **118**, 9149–9156 (2003).
8. G. H. Rawitscher and I. Koltrath, *Comp. in Science and Eng.* **7/6**, 58–66 (2005).
9. T. Kim, and T. Udagawa, *Phys. Rev.* **C42**, 1147–1149 (1990).
10. W. N. Sams, and D. J. Kouri, *J. Chem. Phys.* **51**, 4809–4814 (1969).
11. J. Horáček, and T. Sasakawa, *Phys. Rev.* **A28**, 2151–2156 (1983).
12. J. Horáček, and T. Sasakawa, *Phys. Rev.* **A30**, 2274–2277 (1984)
13. J. Horáček, and T. Sasakawa, *Phys. Rev.* **C32**, 70–75 (1985)
14. J. Horáček, and J. Bok, *Comp. Phys. Comm.* **59**, 319–323 (1990).

A Metaheuristic Supervisory Scheme for Cooperative Manipulators

Maria P. Tzamtzi and Fotis N. Koumboulis

Department of Automation, Halkis Institute of Technology, 34400, Psahna, Evia, Greece

Abstract. A supervisory scheme for cooperative manipulators is proposed, which is used to determine external commands implemented to the cooperative manipulators system through an appropriate control scheme. The supervisor is used to determine external commands representing the desired trajectories of independently controlled state variables that are not related to the desired positioning of the load, thus they may be arbitrary determined so as to contribute towards the achievement of the performance requirements while satisfying configuration, performance and actuator constraints. The external commands are determined as a metaheuristic solution of a multi-inequalities constraints problem.

Keywords: complex systems, metaheuristic algorithm, multi-inequalities constraints, cooperative manipulators, uncertain modeling, supervisory system
PACS: 02.30.Yy, 07.05.Dz, 07.07.Tw, 02.60.Pn, 02.70.-c, 07.05.Mh, 89.20.Bb, 89.20.Kk

INTRODUCTION

Independent control of position and internal forces is of particular importance in multi-manipulator systems (see f.e. [1]-[5]). The use of cooperating manipulators is preferred in several industrial applications instead of the use of a single manipulator, in order to serve complex tasks, where the characteristics of the load and the desired motion, or the requirements for controlling the "squeezing forces" exerted either on the load and/or on the manipulators' links, cannot be adequately served by a single robot. "Squeezing forces" are internal forces that do not affect the motion of the load, however their values are critical, since small values may result to loss of object-manipulator contact, while large values may cause deformation or distortion of the object.

In the present paper, we propose a supervisory scheme for cooperative manipulators used to determine external commands, which represent the desired trajectories of independently controlled state variables and are implemented to the system of cooperative manipulators through an appropriate control scheme. The specific external commands are not related to the desired load's positioning, thus they may be arbitrary determined so as to contribute towards the achievement of performance requirements while satisfying configuration, performance and actuator constraints. The steady state values, as well as the transient performance of the external commands are determined as a metaheuristic solution of a multi-inequalities constraints problem (see f.e. [6]-[8]). The supervisory scheme is presented within the framework of a control design scheme presented in [9], where a linear control law is applied to the multi-manipulator system, while its degrees of freedom are determined with the use of a metaheuristic search algorithm. The metaheuristic determination of the external commands may be incorporated as an enhancement of this algorithm. However, the proposed metaheuristic solution for selecting the external commands is not restricted by the specific control law, and can be equally well combined with other linear or nonlinear control schemes.

DESIGN REQUIREMENTS

Consider a system composed of m manipulators commonly gripping a load, which is either a rigid body or a set of interconnected objects subject to holonomic constraints. Let $q = [q_1^T \cdots q_m^T]^T \in \Box^k$, $\tau = [\tau_1^T \cdots \tau_m^T]^T \in \Box^k$, $f = [f_1^T \cdots f_m^T]^T \in \Box^{6m}$, $k = \sum_{i=1}^m k_i$, $L(p) = \left[L_1^T(p) \cdots L_m^T(p) \right]^T$, $J(q) = \operatorname*{diag}_{i=1,\dots,m} \{ J_i(q_i) \}$, where for the i-th manipulator

CP1046, *Selected Papers from ICNAAM 2007 and ICCMSE 2007*
edited by T. E. Simos, G. Maroulis, G. Psihoyios, and Ch. Tsitouras
© 2008 American Institute of Physics 978-0-7354-0574-5/08/$23.00

$q_i \in \square^{k_i}$ ($k_i \in \square$) are the joint generalized coordinates, τ_i the joint generalized driving forces, $f_i \in \square^6$ the contact forces exerted by the end effector to the carried object, $J_i(q_i) = \partial h_i(q_i)/\partial q_i$ Jacobian matrices, with $h_i(\cdot)$ characterizing the robot's forward kinematics, $L_i(p) = \partial \pi_i(p)/\partial p$ Jacobian matrices, with $p \in \square^l (l \in \square)$ being the generalized coordinates of the load and $\pi_i(\cdot)$ relating the object's position to the position of the i-th robot's end effector with respect to the workspace. Let also f_d denote disturbance forces acting on the load.

The space of forces exerted on the load by the manipulators may be decomposed as follows ([3], [5]): a vector $\overline{f} = L^T(p)f$ of "moving forces" that cause the load's motion and a vector $\hat{f} = S^T(p)f$ of "controllable forces" that may be controlled independently from the load's position. Assuming that $\mathrm{rank}[L(p)] = l$ for all $p \in \square^l$, the matrix $S^T(p) \in \square^{(6m-l) \times 6m}$ is selected so that $[L(p)\ S(p)] \in \square^{6m \times 6m}$ is invertible. This decomposition is described by the equation $[\overline{f}^T\ \hat{f}^T]^T = [L(p)\ S(p)]^T f$.

In [5], the matrix $S(p)$ is appropriately selected so that the controllable forces $\hat{f}$ are projections of the forces exerted on the load, to directions where internal forces, not contributing to the motion of the load, may appear. Using the determination of "squeezing forces" given in [2], we may relate the controllable forces with the "squeezing forces" exerted on the load's objects and/or the manipulators' links with a nonlinear function

$$f_s = f_s(q, \dot{q}, \ddot{q}, p, \dot{p}, \ddot{p}, \hat{f}, \tau, f_d) \tag{1}$$

In [3], a linear static control law has been proposed, that uses measurements of the moving and controllable forces, the load's and the manipulators' generalized variables, as well as their corresponding rates of variation. The linear control law, that does not require knowledge of the load's dynamics, is designed to achieve I/O decoupling for the linearization of the system around an operating point. More specifically, for the linearized system, the load's position coordinates, the controllable forces and a set of $\nu = k - 6m$ of robots' generalized coordinates can be independently controlled via an I/O decoupling scheme, while for all uniquely determined state variables asymptotic stability as well as BIBO stability hold. The application of the linear control law is expected to result in satisfactory performance also for the corresponding nonlinear system, provided that the deviations of the state and input variables from the corresponding operating point remain sufficiently small.

Let $\delta r = [\delta r_1 \cdots \delta r_\nu]$ denote the vector of independently controlled robots' generalized coordinates. Note that δr is a subset of the robots' generalized coordinates $q \in \square^k$, which is selected based on the linear independence properties of the columns of $J(q)$. The selection of δr is not unique and constitutes a degree of freedom for the controller proposed in [3]. Assuming zero initial conditions, the solution of the linear closed-loop system is given by the equations (see [3]):

$$£\{\delta p(t)\} = \operatorname*{diag}_{j=1,\ldots,l}\left\{1/(s^2 + s\mu_{j,1} + \mu_{j,0})\right\} \times \left[\gamma_1 £\{\omega_1(t)\} \cdots \gamma_l £\{\omega_l(t)\}\right]^T \tag{2}$$

$$£\{\delta r(t)\} = \operatorname*{diag}_{j=1,\ldots,\nu}\left\{1/(s^2 + s\lambda_{j,1} + \lambda_{j,0})\right\}\left[\gamma_{l+1} £\{\omega_{l+1}(t)\} \cdots \gamma_{\nu+l} £\{\omega_{\nu+l}(t)\}\right]^T \tag{3}$$

$$£\{\delta \hat{f}(t)\} = \operatorname*{diag}_{j=\nu+l+1,\ldots,k}\{\gamma_j\}\left[£\{\omega_{\nu+l+1}(t)\} \cdots £\{\omega_k(t)\}\right]^T \tag{4}$$

$$£\{\delta \overline{f}(t)\}: \quad \text{not determined} \tag{5}$$

where for each variable, let say z, δz denotes its deviation from the operating point value, $£\{\cdot\}$ denotes Laplace transform, and $\mu_{j,0}$, $\mu_{j,1}$ ($j = 1,\ldots,l$), $\lambda_{j,0}$, $\lambda_{j,1}$ ($j = 1,\ldots,\nu$) and γ_j ($j = 1,\ldots,k$) are arbitrary controller parameters such that $\mu_{j,0} > 0$, $\mu_{j,1} > 0$, $\lambda_{j,0} > 0$, $\lambda_{j,1} > 0$ and $\gamma_j \neq 0$. Finally, $\omega(t) = [\omega_1(t) \cdots \omega_k(t)]^T$ are the external commands, which constitute the desired trajectories for all the independently controlled variables. To achieve the desired positioning of the load, the external commands $\omega_i(t)$ ($i = 1,\ldots,l$) are selected equal to the desired trajectories of the load's coordinates $\delta p(t)$, while the parameters γ_i ($i = 1,\ldots,l$) are selected equal to the corresponding parameters $\mu_{j,0}$, so as to achieve unity steady state gain of the transfer functions $\gamma_j/(s^2 + s\mu_{j,1} + \mu_{j,0})$ in (2). For similar reasons, the parameters γ_i ($i = l+1,\ldots,\nu+l$) are selected equal to the corresponding parameters $\lambda_{j-l,0}$, while the parameters γ_i $(i = \nu + l + 1, \ldots, k)$ are set equal to one. The external commands $\omega_{l+1}(t), \ldots, \omega_{\nu+l}(t)$ and $\omega_{\nu+l+1}(t), \ldots, \omega_k(t)$ represent the desired trajectories for the robots' independently controlled variables $\delta r(t)$ and the controllable forces $\delta \hat{f}(t)$, respectively. As it will be later clarified, $\omega_{l+1}(t), \ldots, \omega_k(t)$ may be appropriately selected by the designer so as to satisfy specific performance requirements.

In [9], the problem of selecting the controller's degrees of freedom has been formulated as an optimization under constraints problem. Then, a metaheuristic algorithm has been proposed, that aims to determine the controller parameters $\mu_{j,0}$, $\mu_{j,1}$ $(j = 1, \ldots, l)$, $\lambda_{j,0}$, $\lambda_{j,1}$ $(j = 1, \ldots, \nu)$, as well as to select the set δr_j $(j = 1, \ldots, \nu)$ of independently controlled robots' generalized coordinates, so as to achieve an optimal, or suboptimal settling time for the load's position p, under the following constraints, determined by the designer based on any available information about the manipulators, the load, the actuators and the task to be performed [9]: 1) Configuration constraints imposed by the structure of the load or the robots and performance constraints imposed by the nature of the specific task to be performed. These constraints concern the variables p, $\dot{p}$, q and $\dot{q}$. 2) Constraints imposed on the deviation of the state and input variables from the nominal operating point, in order to guarantee that the linear controller will be efficient for the nonlinear system. 3) Actuator constraints. 4) Squeezing forces constraints.

Uncertain Modeling of Squeezing Forces

The determination of the nonlinear function f_s in (1), that models the squeezing forces on the load's objects and/or the manipulators' links, may require information related to the geometry of the load's and manipulators' structure and dynamics. However, the designer may select to take under consideration the restrictions concerning only specific "squeezing forces", whose values are critical for the safety and/or the performance of the system. The main performance requirement concerns sufficiently fast positioning of the load, while keeping appropriately bounded the "squeezing forces" exerted on the load, as well as on the manipulators' links. Thus, "squeezing forces" need not to be analytically determined, provided that they are kept within appropriate bounds. Hence, the nonlinear function f_s may describe only some of the squeezing forces, while its determination may be based on partial or even uncertain knowledge of the load's structural and dynamic characteristics. To limit the problems that may be caused due to the uncertainty, the desired bounds of the "squeezing forces" should be determined utilizing appropriate modeling of uncertainties.

An easy to implement approach in many practical applications is to adjust the desired bounds of the "squeezing forces" according to the estimated range of deviations between the real values of the "squeezing forces" and those determined by the function f_s. Estimations of these deviations may be derived utilizing experimental measurements of the forces exerted on the load and the manipulators. For example, if the function f_s is expected to present an error of $\varepsilon\%$, the desired bounds of the "squeezing forces" should be selected $\varepsilon\%$ smaller than their true values. In cases where more rich information is available about the uncertainty, we may use uncertain modeling of "squeezing forces", as for example in the form $|f_s| \leq \left| f_{s,n}(q, \dot{q}, \ddot{q}, p, \dot{p}, \ddot{p}, \hat{f}, \tau, f_d) \right| + b \max\{|q|, |\dot{q}|, |\ddot{q}|, |p|, |\dot{p}|, |\ddot{p}|, |\hat{f}|, |\tau|, |f_d|\}$, where $f_{s,n}$ is a known nominal function and b a known positive constant.

SUPERVISORY SCHEME

The metaheuristic algorithm proposed in [9] for the determination of the controller's degrees of freedom, assumes that the external commands are given by the designer. As already mentioned, the external commands $\omega_i(t)$ $(i = 1, \ldots, l)$ are determined by the task executed by the manipulators, so as to achieve the desired positioning of the load. However, the external commands $\omega_{l+1}(t), \ldots, \omega_{\nu+l}(t)$ that represent the desired trajectories for the robots' independently controlled variables $\delta r(t)$, as well as the external commands $\omega_{\nu+l+1}(t), \ldots, \omega_k(t)$, that represent the desired trajectories for the controllable forces $\delta \hat{f}(t)$, may be arbitrary selected by the designer so as to contribute to the achievement of the performance requirements and the satisfaction of the constraints, presented in Section 2.

Towards this direction, we propose a supervisory scheme, that enhances the functionality of the metaheuristic algorithm used in [9], by incorporating a nested metaheuristic solution of a multi-inequalities constraints problem used for the determination of the external commands. More specifically, for each candidate set of controller parameters, the external commands that control the independently controlled robots' generalized coordinates δr and the controllable forces $\hat{f}$, are determined in a way that the constraint requirements, concerning the robots' generalized coordinates, the joints' generalized forces and the "squeezing forces", are satisfied. The guidelines for the implementation of the metaheuristic solution of this problem, which will be called from now on *ExternalCommandsDesign* algorithm, are presented in the following.

Taking into account that the design goal is to achieve positioning of the carried load, that is carrying the load from an initial to a final desired position, it is obvious that all the state variables, including the robots' generalized coordinates and the "squeezing forces" will result in a corresponding final steady state value. Thus, the *ExternalCommandsDesign* algorithm should determine first a set of steady state values satisfying the imposed constraints. These values are set as the steady state values of the corresponding external commands, while the start values of the external commands are selected equal to the initial conditions of the corresponding variables. The determination of the steady state values is performed with the application of the metaheuristic search algorithm introduced in [9], utilizing computations of the steady state values for the state and input variables of the corresponding closed-loop system.

At the next step, the *ExternalCommandsDesign* algorithm should proceed with the determination of the transition part of each external command, between the initial and the steady state value. For the solution of this problem, we propose the following two approaches:

a) We design the external commands as time functions of a specific form, as for example sigmoid or exponential functions, that satisfy the initial and final conditions. These functions are parameterized by a set of free parameters that determine the settling time and the form of the transition. The *ExternalCommandsDesign* algorithm applies again the metaheuristic search algorithm introduced in [9], in order to determine a set of parameters for the external command functions, such that the resulting closed loop variables of the robotic structure satisfy the constraints. The constraints are tested using simulation results for each candidate external command.

b) Sampling of the external commands is considered, using a sampling period T_e and for a time window $0 \leq t \leq N_e T_e$, determining the transient stage of the external command. For $t \geq N_e T_e$ the external command is set equal to its corresponding steady state value. The values of the external commands for each $\kappa = T_e, 2T_e, \ldots,$ $(N_e - 1)T_e$ may be also determined with the application of the metaheuristic algorithm introduced in [9], searching for each κ a set of values that satisfies the constraints. It is important to note that to avoid discontinuities, the metaheuristic algorithm should search for each instant of time κ in the neighborhood of the values determined for the previous instant of time $\kappa - 1$. When the values of the external commands have been determined for $\kappa = T_e, 2T_e, \ldots, (N_e - 1)T_e$, interpolation may be used to derive the appropriate external command functions.

Before closing this section it is important to note that the previously described metaheuristic solution for the determination of the external commands, may as well be combined with other linear or nonlinear control laws.

ACKNOWLEDGMENTS

The present work is co-financed by the Hellenic Ministry of Education and Religious Affairs' and the ESF of the European Union within the framework of the "Operational Programme for Education and Initial Vocational Training" (Operation "Archimedes-II").

REFERENCES

1. P. Hsu, *IEEE Trans. Robot. and Automat.*, **9**, 400-410 (1993).
2. J.T. Wen, K. Kreutz-Delgado, *Automatica*, **28**, 729-743 (1992)
3. K.G. Tzierakis, F.N. Koumboulis, *J. Franklin Inst.*, **340**, 435-460 (2003).
4. T. Tinós, M.H. Terra, J.Y. Ishihara, *IEEE Trans. on Control Systems Technology*, **14**, 725-734 (2006)
5. M.P. Tzamtzi, F.N. Koumboulis, *IEEE Inter. Conf. on Emerg. Tech. and Factory Autom.*, Czech Republic, 2006, pp. 993-996
6. C. Blum, A. Roli, *ACM Computing Surveys*, **35**, 268–308 (2003).
7. W. Glover, G.A. Kochenberger (Eds.), *Handbook of Metaheuristics*, in Series: *International Series in Operations Research & Management Science*, **57**, 2003, Springer
8. F.N. Koumboulis, M.P. Tzamtzi, "A metaheuristic approach for controller design of multivariable processes", *12th IEEE Int. Conf. Emerging Tech. and Factory Automation, (ETFA 2007)*, Patras, Greece, 2007, to be presented
9. M.P. Tzamtzi, F.N. Koumboulis, "A Metaheuristic Controller for Cooperative Manipulators", *19th International Conference on Tools with Artificial Intelligence*, Patras, Greece, 2007, to be presented

Analysis of the Curvature Tensor from the Viewpoint of Signal Processing

Lennart Wietzke[*], Gerald Sommer[*], Christian Schmaltz[†] and Joachim Weickert[†]

[*]*Institute of Computer Science, Chair of Cognitive Systems, Christian-Albrechts-University of Kiel, Germany*
[†]*Faculty of Mathematics and Computer Science, Mathematical Image Analysis Group, Saarland University, Saarbrücken, Germany*

Abstract. So far the recently introduced monogenic curvature tensor has only been known in Fourier domain. In this paper the monogenic curvature tensor will be formulated in spatial domain as a concatenation of two and three Riesz transforms respectively. Furthermore it will be shown that the Riesz transform of any order can be defined by a concatenation of one dimensional Hilbert transforms in Radon space, the Radon transform of the signal and its inverse. The Riesz, Hilbert and Radon transforms provide a connection between differential geometry and signal processing so that already known results from differential geometry can be used to solve problems from phase based local image analysis of intrinsic dimension two and to interpret the monogenic curvature tensor exactly.

Keywords: Differential Geometry, Signal Processing, Geometric (Clifford) Algebra, Monogenic Curvature Tensor, Monogenic Signal, Radon Transform, Hough Transform, Riesz Transform, Hilbert Transform, Intrinsic Dimension, Analytic Signal, Local Phase Based Image Analysis
PACS: 02.10.Xm,02.40.Dr

INTRODUCTION

In the following 2D signals $f \in \mathbb{R}^\Omega$ will be analyzed in the Monge patch embedding $I_f = \{x\mathbf{e}_1 + y\mathbf{e}_2 + f(x,y)\mathbf{e}_3 | (x,y) \in \Omega \subset \mathbb{R}^2\}$ which is well known from differential geometry with $\{1, \mathbf{e}_1, \mathbf{e}_2, \mathbf{e}_3, \mathbf{e}_{12}, \mathbf{e}_{13}, \mathbf{e}_{23}, \mathbf{e}_{123}\}$ as the set of basis vectors of the Clifford algebra $\mathbb{R}_3$. A 2D signal f will be classified into local regions $N \subseteq \Omega$ of different intrinsic dimension:

$$f \in \begin{cases} \mathrm{i0D}_N, & f(\mathbf{x}_i) = f(\mathbf{x}_j) \quad \forall \mathbf{x}_i, \mathbf{x}_j \in N \\ \mathrm{i1D}_N, & f(x,y) = g(x\cos\theta + y\sin\theta) \quad \forall (x,y) \in N \text{ with } g \in \mathbb{R}^{\mathbb{R}} \text{ and } f \notin \mathrm{i0D}_N \\ \mathrm{i2D}_N, & f(x,y) = \sum_{i \in I} f_i(x,y) \text{ with } f_i \in \mathrm{i1D}_N \text{ and the finite index set } I \text{ with } |I| > 1 \end{cases} \tag{1}$$

The set of signals being analyzed in this work can be written as: $\mathscr{L}_2(\Omega, \mathbb{R}) \cap \bigcup_{N \subseteq \Omega}(\mathrm{i0D}_N \cup \mathrm{i1D}_N \cup \mathrm{i2D}_N)$. The set of i1D signals can be completely analyzed by the monogenic signal [Felsberg] which splits the signal locally into phase, orientation and amplitude information. In this paper the monogenic signal and the monogenic curvature tensor [Zang] will be interpreted in Radon space which gives beautiful access to analyzing Riesz transforms of any order. All odd order Riesz transforms apply a one dimensional Hilbert transform to multidimensional signals in a certain orientation which is determined by the Radon transform [Beyerer]. The Radon transform is defined as:

$$r := r(t,\theta) := \mathscr{R}\{f\}(t,\theta) := \int_{(x,y)\in\Omega} f(x,y)\delta_0(x\cos\theta + y\sin\theta - t)d(x,y) \tag{2}$$

with $\theta \in [0..\pi)$ as the orientation, $t \in \mathbb{R}$ as the minimal distance of the line from the origin and δ_0 as the delta distribution. The inverse Radon transform exists and is defined by:

$$\mathscr{R}^{-1}\{r(t,\theta)\}(x,y) := \frac{1}{2\pi^2}\int_{\theta=0}^{\pi}\int_{t\in\mathbb{R}} \frac{1}{x\cos\theta + y\sin\theta - t}\frac{\partial}{\partial t}r(t,\theta)dt\,d\theta \tag{3}$$

The point of origin where the local phase and orientation information should be obtained within the signal will be translated to position $(0,0)$ for each point $(x,y) \in \Omega$ so that the inverse Radon transform can be simplified to:

$$\mathscr{R}^{-1}\{r\}(0,0) = -\frac{1}{2\pi^2}\sum_{i \in I}\int_{t\in\mathbb{R}} \frac{1}{t}\frac{\partial}{\partial t}r(t,\theta_i)dt \tag{4}$$

CP1046, *Selected Papers from ICNAAM 2007 and ICCMSE 2007*
edited by T. E. Simos, G. Maroulis, G. Psihoyios, and Ch. Tsitouras
© 2008 American Institute of Physics 978-0-7354-0574-5/08/$23.00

because $r(t_1, \theta) = r(t_2, \theta) \; \forall t_1, t_2 \in \mathbb{R} \; \forall \theta \in [0..\pi) - \bigcup_{i \in I}\{\theta_i\}$ implies $\frac{\partial}{\partial t} r(t, \theta) = 0 \; \forall t \in \mathbb{R} \; \forall \theta \in [0..\pi) - \bigcup_{i \in I}\{\theta_i\}$ for a finite number $|I| \in \mathbb{N}$ of superimposed i1D signals which construct the i2D signal $f = \sum_{i \in I} f_i$ where each single i1D signal f_i has its own orientation θ_i.

INTERPRETATION OF THE FIRST ORDER RIESZ TRANSFORM

The Riesz transform of a signal f can be written in terms of the Radon transform:

$$\begin{bmatrix} R_x\{f\}(x,y) \\ R_y\{f\}(x,y) \end{bmatrix} = R\{f\}(x,y) = \mathscr{R}^{-1}\{h_1(t) * r(t, \theta)n_\theta\}(x,y) \tag{5}$$

with $n_\theta = [\cos\theta, \sin\theta]^T$ and h_1 as the one dimensional Hilbert kernel in spatial domain. Proof: Central slice theorem [Felsberg]. Applying the Riesz transform to an i1D signal with orientation θ_m results in:

$$\begin{bmatrix} R_x\{f\}(0,0) \\ R_y\{f\}(0,0) \end{bmatrix} = \underbrace{\left[-\frac{1}{2\pi^2} \int_{t \in \mathbb{R}} \frac{1}{t} h_1(t) * \frac{\partial}{\partial t} r(t, \theta_m) dt \right]}_{=:s(\theta_m)} n_{\theta_m} \tag{6}$$

Note that $\frac{\partial}{\partial t}(h_1(t) * r(t, \theta)) = h_1(t) * \frac{\partial}{\partial t} r(t, \theta)$. The orientation of the signal can therefore be derived by $\arctan \frac{R_y\{f\}(0,0)}{R_x\{f\}(0,0)} = \arctan \frac{s(\theta_m)\sin\theta_m}{s(\theta_m)\cos\theta_m} = \theta_m$ (see also figure 1). The Hilbert transform of f and with it also the one dimensional phase can be calculated by: $\sqrt{[R_x\{f\}(0,0)]^2 + [R_y\{f\}(0,0)]^2} = h_1 * f_{\theta_m}$ with the partial Hilbert transform $(h_1 * f_\theta)(0) = -\frac{1}{\pi} \int_{\tau \in \mathbb{R}} \frac{f(\tau\cos\theta, \tau\sin\theta)}{\tau} d\tau$. The first order Riesz transform of any i2D signal consisting of a number

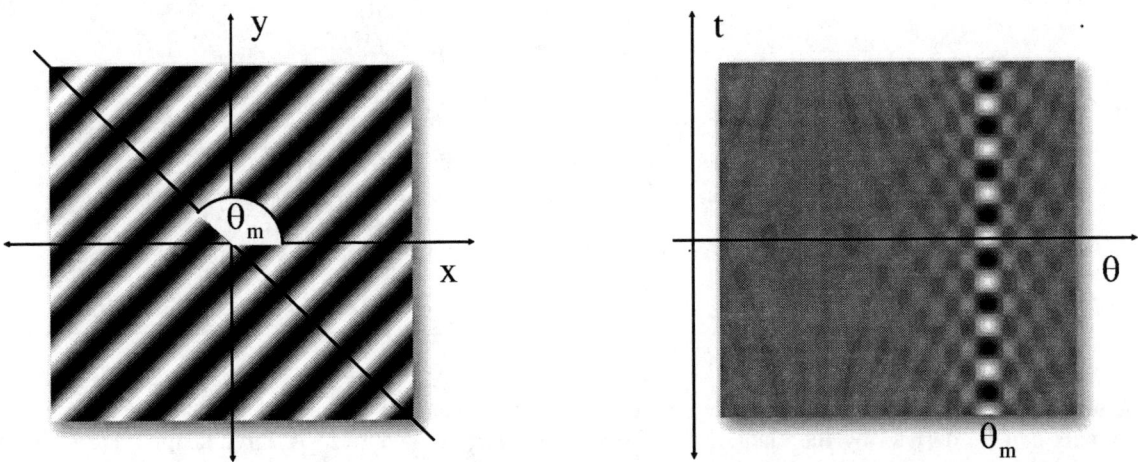

FIGURE 1. From left to right: i1D signal in spatial domain with x- and y-axis and with main orientation θ_m and in Radon space with θ- and t-axis. Even though in the right figure some artifacts can be seen in the Radon space, those artifacts do not exist in case of an infinite support of the image patterns and therefore the artifacts will be neglected in this work: $\frac{\partial}{\partial t} r(t, \theta) = 0 \; \forall t \; \forall \theta \neq \theta_m$ and in the left figure $r(t_1, \theta) = r(t_2, \theta) \; \forall t_1, t_2 \; \forall \theta \neq \theta_m$

$|I|$ of i1D signals reads:

$$\begin{bmatrix} R_x\{f\}(0,0) \\ R_y\{f\}(0,0) \end{bmatrix} = \sum_{i \in I} \left[-\frac{1}{2\pi^2} \int_{t \in \mathbb{R}} \frac{1}{t} h_1(t) * \frac{\partial}{\partial t} r(t, \theta_i) dt \right] n_{\theta_i} = \sum_{i \in I} s(\theta_i) n_{\theta_i} \tag{7}$$

THE MONOGENIC CURVATURE TENSOR

So far the monogenic curvature tensor [Zang] has only been known in Fourier domain. This drawback makes interpretation impossible when applying the monogenic curvature tensor to a certain signal model. This problem can be solved in spatial domain of the Riesz transform. Now any i2D signal will be regarded as the superposition of a finite number of i1D signals $f = \sum_{i \in I} f_i$. With the properties $\mathcal{R}\{\mathcal{R}^{-1}\{r\}\} = r$ and $\mathcal{R}\{\sum_{i \in I} f_i\} = \sum_{i \in I} \mathcal{R}\{f_i\}$ [Toft] and because of $h_1 * h_1 * f = -f$ the monogenic curvature tensor $T = T_e + T_o \in M(2, \mathbb{R}_3)$ can be also written in terms of the concatenation of two and three Riesz transforms respectively and therefore also in terms of the Radon transform and its inverse [Stein]:

$$T_e(f) = \mathcal{F}^{-1}\left\{ \mathcal{F}\{f\}(\alpha, \rho) \begin{bmatrix} \cos^2 \alpha & -\sin \alpha \cos \alpha \, \mathbf{e}_{12} \\ \sin \alpha \cos \alpha \, \mathbf{e}_{12} & \sin^2 \alpha \end{bmatrix} \right\} = \begin{bmatrix} R_x\{R_x\{f\}\} & -R_x\{R_y\{f\}\}\mathbf{e}_{12} \\ R_x\{R_y\{f\}\}\mathbf{e}_{12} & R_y\{R_y\{f\}\} \end{bmatrix} \quad (8)$$

With $\mathcal{F}\{f\}(\alpha, \rho)$ as the Fourier transformed of the signal f in polar coordinates with α as the angular component and ρ as the radial component and $\mathcal{F}^{-1}$ as the inverse Fourier transform. Now the even tensor in Radon space reads:

$$T_e(f) = -\begin{bmatrix} \mathcal{R}^{-1}\{\cos^2 \theta \mathcal{R}\{f\}\} & -\mathcal{R}^{-1}\{\sin \theta \cos \theta \mathcal{R}\{f\}\}\mathbf{e}_{12} \\ \mathcal{R}^{-1}\{\sin \theta \cos \theta \mathcal{R}\{f\}\}\mathbf{e}_{12} & \mathcal{R}^{-1}\{\sin^2 \theta \mathcal{R}\{f\}\} \end{bmatrix} \quad (9)$$

The odd tensor is defined as the Riesz transform of the even part: $T_o(f) = T_e(R_x\{f\} + R_y\{f\}\mathbf{e}_{12})$. Please note that the monogenic curvature tensor and even the monogenic signal are not restricted to any intrinsic dimension.

Interpretation of the Monogenic Curvature Tensor for Two Superimposed i1D Signals

In the following two superimposed i1D signals with orientations θ_1, θ_2 and arbitrary but same phase φ for both i1D signals will be analyzed in Radon space (see figure 2). With the abbreviations: $a := \cos \theta_1$, $b := \cos \theta_2$, $c := \sin \theta_1$, $d := \sin \theta_2$ the even tensor for two i1D signals reads:

$$T_e = f(0,0) \begin{bmatrix} a^2 + b^2 & -(ca + db)\mathbf{e}_{12} \\ (ca + db)\mathbf{e}_{12} & c^2 + d^2 \end{bmatrix} \quad (10)$$

and the odd tensor for two i1D signals reads:

$$T_o = -s \begin{bmatrix} (a^3 + b^3) + (ca^2 + db^2)\mathbf{e}_{12} & -(ca^2 + db^2)\mathbf{e}_{12} + (c^2 a + d^2 b) \\ (ca^2 + db^2)\mathbf{e}_{12} - (c^2 a + d^2 b) & (c^2 a + d^2 b) + (c^3 + d^3)\mathbf{e}_{12} \end{bmatrix} \quad (11)$$

with assumption $s := s(\theta_1) = s(\theta_2)$. The determinant of the odd tensor reads:

$$\mathbf{e}_1 \det T_o = \mathbf{e}_1 \underbrace{s^2(a^3 d^2 b + b^3 c^2 a - ca^2 d^3 - db^2 c^3 - 2ca^2 db^2 + 2c^2 ad^2 b)}_{=:B} + \mathbf{e}_2 \underbrace{s^2(a^3 d^3 + b^3 c^3 - db^2 c^2 a - ca^2 d^2 b)}_{=:C} \quad (12)$$

Using that notation, the main orientation can be derived by: $\theta_m = \frac{\theta_1 + \theta_2}{2} = \frac{1}{2}\arctan \frac{C}{B}$. The apex angle of the two i1D signals can be derived by the Gaussian and main curvature known from differential geometry and with the basic ideas of Euler's and Meusnier's theorems [Baer]: $H := \frac{1}{2}\text{trace}(T_e) = \frac{1}{2}(a^2 + b^2 + c^2 + d^2)f(0,0) = f(0,0)$ and $K := \det T_e = [f(0,0)]^2(a^2 d^2 + b^2 c^2 - 2abcd)$. This yields the apex angle:

$$\left| \frac{\theta_1 - \theta_2}{2} \right| = \arctan \sqrt{\left| \frac{H - \sqrt{H^2 - K}}{H + \sqrt{H^2 - K}} \right|} = \arctan \sqrt{\left| \frac{1 - \sqrt{1 - a^2 d^2 - b^2 c^2 + 2abcd}}{1 + \sqrt{1 - a^2 d^2 - b^2 c^2 + 2abcd}} \right|} \quad (13)$$

The phase φ of both signals can be derived in many different ways. One possibility would be by the determinants of the even and odd tensors: $\varphi = \arctan \frac{B^2 + C^2}{\det^2 T_e}$.

CONCLUSION AND FUTURE WORK

The odd-order Riesz transform of any i1D or i2D signal can be analyzed in Radon space in which a one dimensional partial Hilbert transform will be applied in direction of each i1D signal with its individual orientation θ_i, $i \in I$.

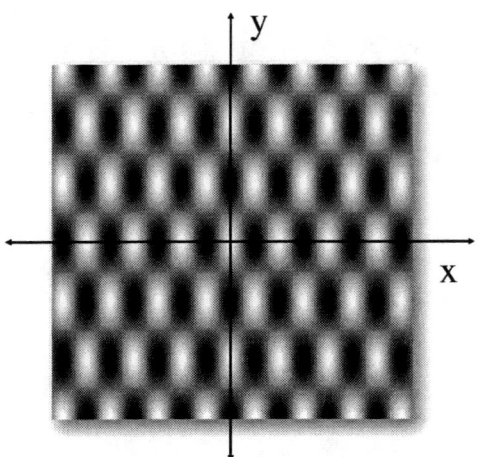 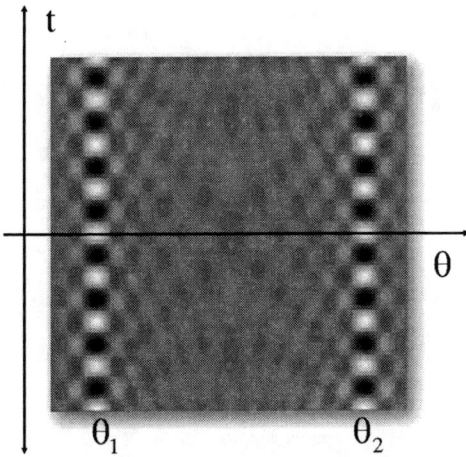

FIGURE 2. From left to right: Two superimposed i1D signals with orientation θ_1 and θ_2 in spatial domain and in Radon space. Assuming that both signals have same phase φ the signal information can be separated by the Riesz transform.

Assuming that two superimposing i1D signals have arbitrary but same phase φ, the orientations θ_1, θ_2 and the phase can be calculated. The advantage of analyzing the Riesz transform in Radon space is that the signal properties consisting of orientation and phase are explicitly given after applying the general operator (e.g. the monogenic curvature operator) to the specific i1D or i2D signal function. Future work will include analyzing the superposition of i1D signals with individual phases φ_i. Assume the orientations θ_1, θ_2 of the two signals to be known. Then the Hilbert transforms of each signal can be derived by the following linear system of equations:

$$\begin{bmatrix} h_1 * f_{\theta_1} \\ h_1 * f_{\theta_2} \end{bmatrix} = \frac{1}{ad - cb} \begin{bmatrix} R_x\{f\}d - R_y\{f\}b \\ R_y\{f\}a - R_x\{f\}c \end{bmatrix} \tag{14}$$

Note that arbitrary but same phase of both signals has been assumed in this paper for deriving the orientations, so that only one common phase can be calculated by this assumption. The analysis of i2D signals in Radon space presented in this work realizes to interpret the monogenic curvature tensor for the first time in an exact way.

ACKNOWLEDGMENTS

We acknowledge funding by the German Research Foundation (DFG) under the projects *SO 320/4-2* and *We 2602/5-1*.

REFERENCES

Beyerer.   Jürgen Beyerer and Fernando Puente Leon, The Radon Transform in Digital Image Processing, Publisher: Oldenbourg, 2002, Vol 50, Part 10, pages: 472-480, Country of publication: Germany, Journal title: Automatisierungstechnik, ISSN: 0178-2312

Toft.   Peter Toft, The Radon Transform - Theory and Implementation, PhD Thesis, 1996, Department of Mathematical Modelling, Technical University of Denmark, June 1996

Baer.   Christian Baer, Elementare Differentialgeometrie, de Gruyter, 2001, Vol 1, ISBN-10: 3110155206, ISBN-13: 978-3110155204

Felsberg.   Michael Felsberg, Low-Level Image Processing with the Structure Multivector, PhD Thesis, Kiel University, 2002

Zang.   Di Zang, Signal Modeling for Two-Dimensional Image Structures and Scale-Space Based Image Analysis, PhD Thesis, Kiel University, 2007

Stein.   Elias M. Stein, Singular integrals and differentiability properties of functions, Princeton Mathematical Series, Princeton University Press, Princeton, New Jersey, 1970

Periodic Solutions of 2D Nonlinear Wave Equations in Rectangle Domains

Masaru Yamaguchi

Department of Mathematics, Tokai University, 259-1292 Hiratsuka, Kanagawa, Japan

Abstract. We shall show the existence of infinitely many time-periodic solutions of BVP for a 2D nonlinear wave equation defined in a rectangle domain.

Keywords: periodic solutions,nonlinear wave equation,rectangle domain
PACS: 03.65.Ge,03.65.Db,02.30.Jr

INTRODUCTION

Let B be a rectangular domain

$$B : 0 < x_1 < \frac{\pi}{a}, \quad 0 < x_2 < \frac{\pi}{b},$$

where a and b are positive constants. Let D be a rectangular cylindrical domain $D = B \times R_t^1$. Let $\triangle$ be the 2-dimensional Laplacian $\partial_{x_1}^2 + \partial_{x_2}^2$ and $x = (x_1, x_2)$.

We shall deal with BVP to a nonlinear wave equation

$$\begin{cases} \partial_t^2 u(x,t) - \triangle u(x,t) + f(x, u(x,t)) = 0, & (x,t) \in D, \\ u(x,t) = 0, & (x,t) \in \partial D. \end{cases} \tag{P}$$

Here we assume the following condition on f and a, b.

(A) $f(x, u)$ is of C^∞-class in $(x, u) \in B \times R_u^1$ and odd in u, and satisfies

$$f(x, 0) = \partial_u f(x, 0) = \partial_u^2 f(x, 0) = 0$$

for $x \in B$.

(B) There exists a constant $\tilde{\omega} > 0$ such that

$$|k^2 \tilde{\omega}^2 - (j_1^2 a^2 + j_2^2 b^2)| \geq C_0$$

for all $(k, j_1, j_2) \in Z \times N^2$ and some constant $C_0 > 0$.

Our main purpose of this paper is to show the the following theorem and to clarify the structure of those periodic solutions.

Theorem 1. *Assume* (A) *and* (B). *Then* BVP (P) *has infinitely many time-periodic solutions of C^2-class.*

This theorem is the direct conclusion of Theorem 2 in section 2. To show Theorem 2 we shall apply the Lyapunov center theorem of infinite dimension developed by Bambusi [Ba] and the number-theoretic property of the periods and a, b studied by [Ya2].

Consider the linear BVP

$$\begin{cases} \partial_t^2 u(x,t) - \triangle u(x,t) = 0, & (x,t) \in D, \\ u(x,t) = 0, & (x,t) \in \partial D. \end{cases} \tag{LP-a}$$

We shall look for infinitely many periodic solutions of (P) near every normal mode of (LP-a) of the form

$$c \cos q\omega_0 t \sin p_1 a x_1 \sin p_2 b x_2 \tag{1}$$

CP1046, *Selected Papers from ICNAAM 2007 and ICCMSE 2007*
edited by T. E. Simos, G. Maroulis, G. Psihoyios, and Ch. Tsitouras
© 2008 American Institute of Physics 978-0-7354-0574-5/08/$23.00

for some $q \in Z_+$, $(p_1, p_2) \in N \times N$ and $\omega_0 > 0$ satisfying

$$q^2 \omega_0^2 = p_1^2 a^2 + p_2^2 b^2. \tag{2}$$

Here c is a small constant.

We give notation and definitions. Let Z_+ be ths set of nongegative integers. Let O be any open set in R^n and s be in Z_+. $L^2(O)$ and $H^s(O)$, $H_0^s(O)$ are the usual Lebesgue and Sobolev spaces respectively. We set $H^0(O) = L^2(O)$. Let T be a positive number. We denote $B \times (0, T)$ by Ω_T and $B \times R_t^1$ by Ω. Let the norms of $L^2(\Omega_T)$ and $H^s(\Omega_T)$ be denoted by $|\cdot|_{0,T}$ and $|\cdot|_{s,T}$. $C_{per,T,e}^\infty(\Omega)$ is the space whose elements belong to $C^\infty(\Omega)$ and are T-periodic and even in t. $C_{0,per,T,e}^\infty(\Omega)$ is the subspace of $C_{per,T,e}^\infty(\Omega)$ whose elements has the supports in Ω. $H_{per,T,e}^s(\Omega)$ and $H_{0,per,T,e}^s(\Omega)$ are the completions with respect to $|\cdot|_{s,T}$ of $C_{per,T,e}^\infty(\Omega)$ and $C_{0,per,T,e}^\infty(\Omega)$ respectively. $H_{per,T,e}^s(\Omega)$ and $H_{0,per,T,e}^s(\Omega)$ are identified with the subspaces of $H^s(\Omega_T)$ and $H_0^s(\Omega_T)$. When $T = 2\pi$, we set $\Pi = \Omega_{2\pi}$ and $|\cdot|_s = |\cdot|_{s,2\pi}$. We set $H_{per,e}^s(\Omega) = H_{per,2\pi,e}^s(\Omega)$ and $H_{0,per,e}^s(\Omega) = H_{0,per,2\pi,e}^s(\Omega)$ for simplicity. Clearly $H_{per,e}^0(\Omega) = L_{per,e}^2(\Omega)$.

PERIODIC SOLUTIONS OF (P)

We shall rewrite (P) to the following BVP by changing time variable as $\theta = \omega t$, where $\omega > 0$ is a parameter, and unknown function as $v(x, \theta) = u(x, \theta/\omega)$

$$\begin{cases} \omega^2 \partial_\theta^2 v - \triangle v + f(x, v) = 0, & (x, \theta) \in \Omega, \\ v(x, \theta) = 0, & (x, \theta) \in \partial\Omega. \end{cases} \tag{MP}$$

We shall look for 2π-periodic solutions of (MP) for ω in a suitable set S with infinitely many elements. Then (P) has infinitely many $2\pi/\omega$-periodic solutions with $\omega \in S$.

We take the base space for BVP (MP) as $X_s = H_{per,e}^s(\Omega) \cap H_{0,per,e}^1(\Omega)$.

Decomposition of BVP (MP)

We shall obtain periodic solutions near normal modes. To this end we decompose (P) into a problem for normal mode direction and a problem for its orthogonal direction. Let

$$e_k(\theta) = \frac{1}{\sqrt{\pi}} \cos k\theta, \quad \psi_j(x) = \frac{2\sqrt{ab}}{\pi} \sin j_1 a x_1 \sin j_2 b x_2,$$

$$\phi_{jk}(x, \theta) = e_k(\theta) \psi_j(x),$$

where $j = (j_1, j_2)$.

Let $p \in N \times N$, $q \in Z_+$ and ω_0 be the same constants as in (2). Let $M = M(p, q; \omega_0)$ be the subspace of $L_{per,e}^2(\Omega)$ spanned by $\phi_{pq}(x, \theta; \omega_0)$ written as $\phi(x, t)$ for simplicity. Let $M^\perp = M(p, q; \omega_0)^\perp$ be the orthogonal complement of M in $L_{per,e}^2(\Omega)$. We denote the projectors of $L_{per,e}^2(\Omega)$ onto M and $M^\perp$ by P and $P^\perp$ respectively. We decompose the unknown function v into the direct sum in (MP)

$$v = \varepsilon \phi + \varepsilon^\alpha w, \tag{3}$$

where $w \in M^\perp$ and $\alpha > 0$ is a constant with $0 < \alpha < 1$. Then operating P and $P^\perp$ to BVP (MP), we obtain

$$-q^2(\omega^2 - \omega_0^2)\phi + \varepsilon^{-1} P f(x, \varepsilon \phi + \varepsilon^\alpha w) = 0, \quad (x, \theta) \in \Omega, \tag{A}$$

where ω_0 is the same constant as in (2), and

$$\begin{cases} \omega^2 \partial_\theta^2 w - \triangle w + \varepsilon^{-\alpha} P^\perp f(x, \varepsilon \phi + \varepsilon^\alpha w) = 0, & (x, \theta) \in \Omega, \\ w(x, \theta) = 0, & (x, \theta) \in \partial\Omega. \end{cases} \tag{B}$$

We recall the following result on the periodic eigenvalue problem

$$\begin{cases} (\omega^2 \partial_\theta^2 - \triangle)\rho(x,\theta) = \lambda\,\rho(x,\theta), & (x,\theta) \in \Omega, \\ \rho(x,\theta) = 0, & (x,\theta) \in \partial\Omega, \\ \rho(x,\theta+2\pi) = \rho(x,\theta), & (x,\theta) \in \Omega. \end{cases} \qquad \text{(EVP)}$$

We know that (EVP) has the eigenvalues $\lambda_{jk} = j_1^2 a^2 + j_2^2 b^2 - k^2 \omega^2$ and the corresponding eigenfunctions ϕ_{jk}. $\{\phi_{jk}\}$ is a CONS in $L^2_{per,e}(\Omega)$, and a complete and orthogonal system in $H^1_{0,per,e}(\Omega)$.

Number-Theoretic Conditions

We shall consider the number theoretic condition (B) on a, b and ω, i.e., the Diophantine condition, in order to describe properties of a, b, ω for which (P) has periodic solutions. This clarifies some number theoretic property of the periods of solutions (See Proppositions 1 and 2).

We have the following propositions ([Ya2]) that assure the existence of infinitely many ω satisfying (B) for any fixed relatively prime integers a^2 and b^2 and for $a = b = 1$ respectively. Clearly we can take $C_0 = 1$ in (NC) in both cases.

Proposition 1 ([Ya2]). *Let A and B be any relatively prime positive integers. Then there exist infinitely many positive prime integers C such that the Diophantine equation $A i^2 + B j^2 = C k^2$ has no nontrivial integral solution (i,j,k).*

Proposition 2 ([Ya2]). *Assume that $C \in N$ is a prime number and satisfies $C \equiv 3 \mod 4$. Then the Diophantine equation $i^2 + j^2 = C k^2$ has no nontrivial integral solution (i,j,k).*

For the proofs of the above propositions, see [Ya2].

Periodic Solutions of BVP (MP)

Theorem 2. *Assume (A) and (B). Then there exists a sequence $\{(\omega_i, \varepsilon_i)\}_{i \geq 1}$ such that BVP (MP) with $\omega = \omega_i$ has a 2π-periodic solution $v \in X_s$ of the form (3), where $w \in X_s \cap M^\perp$ is the solution of BVP (B) with $\omega = \omega_i$ and $\varepsilon = \varepsilon_i$.*

First we shall solve BVP (B) for any fixed ω satisfying (NC). To this end we shall consider the following linear BVP

$$\begin{cases} \omega^2 \partial_\theta^2 u(x,\theta) - \triangle u(x,\theta) = g(x,\theta), & (x,\theta) \in \Omega, \\ u(x,\theta) = 0, & (x,\theta) \in \partial\Omega, \end{cases} \qquad \text{(LP-b)}$$

where $g(x,\theta)$ is 2π-periodic and even in θ.

The following fundamental propositions show the existence and uniqueness theorem of a 2π-periodic solution with estimate of the solution for (LP-b). They are shown in the same way as in [Ya2].

Proposition 3. *Assume that ω satisfies (B). Let $g \in M^\perp$. Then (LP-b) has a unique 2π-periodic solution u in $M^\perp$. u satisfies*

$$|u|_0 \leq C_0^{-1} |g|_0, \qquad (4)$$

where C_0 is the same constant as in (NC).

Proposition 4. *Let $s \geq 1$. Assume that ω satisfies (B). Let $g \in X_s \cap M^\perp$. Then (LP-b) has a unique 2π-periodic soleution u in $X_s \cap M^\perp$. u satisfies*

$$|u|_s \leq C C_0^{-1} |g|_s, \qquad (5)$$

where C is a constant dependent on a, b and independent of ω.

We shall deal with BVP (B)

$$\begin{cases} \omega^2 \partial_\theta^2 w - \triangle w + \varepsilon^{-\alpha} P^\perp f(x, \varepsilon\phi + \varepsilon^\alpha w) = 0, & (x, \theta) \in \Omega, \\ w(x, \theta) = 0, & (x, \theta) \in \partial\Omega. \end{cases} \tag{LP-c}$$

Based on the above propositions, we show the following proposition.

Proposition 5. *Let $s \geq 4$ and r be any positive number. Assume (A) and (B). Then there exists $\varepsilon_0 > 0$ dependent on C_0, r such that BVP (B) has a solution w in $X_s \cap M^\perp$ for any ε with $|\varepsilon| \leq \varepsilon_0$. w has the estimate $|w|_s \leq r$.*

Remark 1. ε_0 is of $O(C_0)$ $(C_0 \to 0)$.

Proof of Proposition 5. Define a nonlinear operator $A = A_{\varepsilon,\omega}$ by

$$Aw = (\omega^2 \partial_\theta^2 - \triangle)^{-1} \varepsilon^{-\alpha} P^\perp f(x, \varepsilon\phi + \varepsilon^\alpha w).$$

Let r be any positive number and $G_r = \{f \in X_s; |f|_s \leq r\}$.

We shall show that there exists $\varepsilon_0 = \varepsilon_0(C_0, r) > 0$ such that $A_{\varepsilon,\omega}$ maps G_r into G_r and is contracting for any ε with $|\varepsilon| \leq \varepsilon_0$. Let $\bar{C}_i > 0$ be constants dependent on the Sobolev constants, ϕ, r, f and its derivatives of up to order $s+1$ and independent of ω and ε. Let $\bar{C}_i$ be positive constants dependent on r and f and its derivatives. Let $w \in G_r \cap M^\perp$. From Proposition 4 we see

$$|Aw|_s \leq C C_0^{-1} |\varepsilon^{-\alpha} P^\perp f(x, \varepsilon\phi + \varepsilon^\alpha w)|_s.$$

Then by the Sobolev inequality we have $|w|_{L^\infty(\Omega)} \leq C(s)|w|_s \leq C(s)r$ for $s \geq 4$, where $C(s)$ is the Sobolev constant. Then using the Moser inequality in [Ya1] and the fact that X_s is a Banach algebra, we have

$$|Aw|_s \leq \varepsilon^{-\alpha} \bar{C}_1 C_0^{-1} (|\varepsilon\phi + \varepsilon^\alpha w|_s^3 (|\varepsilon\phi + \varepsilon^\alpha w|_s + 1),$$
$$\leq \varepsilon^{2\alpha} \bar{C}_2 C_0^{-1} (|\phi|_s + r)^3 (|\phi|_s + r + 1).$$

Hence we obtain

$$|Aw|_s \leq \bar{C}_3 (\varepsilon^{2\alpha} C_0^{-1}). \tag{6}$$

In the similar way we obtain

$$|Aw_1 - Aw_2|_s \leq \bar{C}_4 (\varepsilon^{2\alpha} C_0^{-1})|w_1 - w_2|_s \tag{7}$$

for $w_1, w_2 \in G_r \cap M^\perp$. Therefore take ε_0 so as to satisfy $\bar{C}_3 \varepsilon_0^{2\alpha} C_0^{-1} \leq r$ and $\bar{C}_4 \varepsilon_0^{2\alpha} C_0^{-1} \leq 1/2$. Then it follows that for any ε with $|\varepsilon| \leq \varepsilon_0$ A is a contraction mapping of G_r into itself. Then by the contraction mapping principle we obtain the fixed point w of A unique in G_r. Clearly w is a solution of BVP (B) in $X_s \cap M^\perp$. w is unique in G_r. $\square$

Let $\omega_0 > 0$ be the minimum number which satisfies (2). We look for periodic solutions near a normal mode (1).

Lemma 1. *There exist infinitely many pairs $\{(\varepsilon_i, \omega_i)\}_{i \geq 1}$ contained in $(-\varepsilon_0, \varepsilon_0) \times R_+$ which satisfy Eq. (A). Each ω_i satisfies (B).*

Proof. Let $\tilde{\omega} > 0$ satisfy (B) with $\tilde{C}_0$ as the constant C_0 in (B). Take a sequence of convergents r_i/s_i of

$$\omega_0/\tilde{\omega} = [a_0; a_1, a_2, \cdots],$$

where the right hand side is the continued fraction expansion. Then it follows from [Kh] that

$$\frac{\tilde{\omega}}{s_i^2(a_{i+1} + 2)} \leq \left|\omega_0 - \frac{r_i}{s_i}\tilde{\omega}\right| \leq \frac{\tilde{\omega}}{s_i^2 a_{i+1}}. \tag{8}$$

Note that the right hand side tends to 0 as $i \to \infty$. We set $\omega_i = (r_i/s_i)\tilde{\omega}$. ω_i satisfies (B). In fact, since $\tilde{\omega}$ satisfies (B), it follows that

$$|k^2 \omega_i^2 - (j_1^2 a^2 + j_2^2 b^2)| = \frac{1}{s_i^2}|k^2 r_i^2 \tilde{\omega}^2 - s_i^2(j_1^2 a^2 + j_2^2 b^2)| \geq \frac{\tilde{C}_0}{s_i^2}. \tag{9}$$

From Eq. (A) ω_i and ε satisfy

$$\omega_i^2 - \omega_0^2 = \frac{1}{\varepsilon q^2} (f(x, \varepsilon \phi + \varepsilon^\alpha w), \phi)_{L^2}. \tag{10}$$

Since $f(x, u)$ is of order 3 with respect to u, the right habd side of (10) is of order $3\alpha - 1$ with respect to ε. Hence by the implicit function theorem it follows that there exists ε_i satisfying (10). $\qquad\square$

Remark 2. In the above proof it follows from (8) that

$$\varepsilon_i \sim (s_i^2 a_{i+1})^{-1/(3\alpha-1)} \quad (i \to \infty). \tag{11}$$

Now Theorem 2 is proved without difficulty. From (9) the constant for ω_i in (9) is $\tilde{C}_0/s_i^2$. It follows from Remark 2 that the right hand sides of (6) and (7) are of order

$$\varepsilon_i^{2\alpha} s_i^2 \sim s_i^{2 - \frac{4\alpha}{3\alpha-1}} a_{i+1}^{-\frac{2\alpha}{3\alpha-1}}.$$

Taking α as $1/3 < \alpha < 1$ the above exponent of s_i is negative. Hence if we take i_0 suitably large, then for any i with $i \geq i_0$.

$$\begin{cases} A w \in G_r, \\ |Aw_1 - Aw_2|_s \leq (1/2)|w_1 - w_2|_s. \end{cases}$$

This shows that Eq. (A) and BVP (B) are solved for all $(\omega_i, \varepsilon_i)$, $i \geq i_0$. By reordering i from $i = 1$, we obtain the conclusion of Theorem 2.

REFERENCES

Ba. D. Bambusi, Lyapunov center theorem for some nonlinear PDE's : A simple proof, *Aan. Scuola Norm. Sup. Pisa CL. Sci.* (4), Vol.29 (2000), 823-837.

Kh. A. Ya. Khinchin, *Continued Fractions,* The University of Chicago Press, 1964.

Ya1. M. Yamaguchi, Existence and stability of global bounded classical solutions of initial boundary value problem for semilinear wave equations, *Funkcialaj Ekvacioj*, Vol. 23 (1980), 289-308.

Ya2. M. Yamaguchi, Existence of periodic solutions of second order nonlinear evolution equations and applications, *Funkcialaj Ekvacioj*, Vol. 38 (1995), No. 3, 519-538.

Ya3. M. Yamaguchi, Free and forced vibrations of nonlinear wave equations in ball, *J. Differential Equations*, Vol. 203 (2004), No. 1, 255-291.

Olivier, Y., 32

P

Pantos-Kikkos, S., 110
Papakonstantinou, K., 110
Park, D.-W., 99
Penzov, A. A., 114
Perreault, F., 52
Petelenz, P., 36
Phung, T. V. B., 64
Pop, I., 119
Puścian, P., 133

R

Revnic, C., 119

S

Sakaki, S., 44
Samko, N., 126
Sasagane, K., 40
Sato, H., 44
Schmaltz, C., 150
Shigeta, Y., 48
Sobieski, S., 130, 133
Soldera, A., 52
Sommer, G., 95, 150
Stoilova, S. S., 114
Sulaiman, W. T., 135
Šulc, M., 142
Suzuki, M., 68

T

Takada, T., 19
Takahashi, H., 15, 56
Takano, Y., 60
Takebe, A., 15
Tarana, M., 142
Tzamtzi, M. P., 102, 146

U

Ukai, T., 19

V

van Duin, A. C. T., 23
van Santen, R., 23

W

Weickert, J., 150
Wietzke, L., 150

Y

Yamaguchi, K., 19, 28
Yamaguchi, M., 154
Yamaki, D., 68
Yamanaka, S., 19
Yokogawa, D., 44

Z

Zannoni, C., 32